Vinod Wadhawan

Latent, Manifest, and Broken Symmetry

BOOKS BY VINOD WADHAWAN

Introduction to Ferroic Materials
Gordon & Breach Science Publishers, Amsterdam (2000)
ISBN 90-5699-286-4

Smart Structures: Blurring the Distinction between the Living and the Nonliving
Oxford University Press, Oxford (2007)
ISBN 978-0-19-922917-8

Complexity Science: Tackling the Difficult Questions We Ask about Ourselves and about Our Universe
LAP Lambert Academic Publishing, Saarbrücken (2010)
ISBN 978-3-8383-7754-4

Nauka Zlozonosci: Trudne pytania, ktore zadajemy o sobie i o naszym Wszechswiecie
(In Polish; translation by Malgorzata Koraszewska)
Racjonalista.pl, Wroclaw (2010)
ISBN 978-83-62503-02-5

Latent, Manifest, and Broken Symmetry: A Bottom-up Approach to Symmetry, with Implications for Complex Networks
CreateSpace Independent Publishing Platform, SC, USA (2011/2018)
ISBN 978-1463766719

Understanding Natural Phenomena: Self-Organization and Emergence in Complex Systems
CreateSpace Independent Publishing Platform, SC, USA (2017/2018)
ISBN 978-1548527938

Art forms of Nature. Radial symmetry in sea anemones.
From Ernst Haeckel's *Kunstformen der Natur*
http://commons.wikimedia.org/wiki/File:Haeckel_Actiniae.png

Latent, Manifest, and Broken Symmetry

A Bottom-up Approach to Symmetry, with Implications for Complex Networks

Vinod Wadhawan

CreateSpace Independent Publishing Platform, SC, USA
2018

For my grandchildren
Rhea, Richa, Nishka, Arjun

Long before any science, man was fascinated by symmetry.
Klaus Mainzer

The biggest conceptual change over the last 100 years in the way
physicists think about the world is symmetry.
Lawrence Krauss

First edition published in 2011
Reprinted with corrections: 2014
Second edition: 2018

Contents

Foreword   xi

Preface   xiii

1. Overview   1

2. Symmetry Fundamentals   9
2.1   Definition of symmetry 9
2.2   Analogy and classification are symmetry 11
2.3   Reduction is symmetry 11
2.4   Reproducibility is symmetry 13
2.5   Predictability is symmetry 14
2.6   The symmetry principle 15
2.7   Thermodynamics and the symmetry principle 16
2.8   Ugly symmetry 17

3. Group-Theoretical Description of Symmetry   21
3.1   Discrete groups 21
3.2   Coset decomposition of a group 23
3.3   Lagrange theorem for subgroups 25
3.4   Symmetry group of a crystal 25
3.5   Continuous groups 27
3.6   Permutation groups 27
3.7   Special unitary groups 27
3.8   Topological space, open sets 28
3.9   Morphisms, categories 29
3.10   Semigroups, groupoids 30
3.11   Lie groups 32

4. Network Theory   39
4.1   Mathematical networks 39
4.2   Clustering coefficient 42
4.3   Permutation symmetry in graphs and networks 43
4.4   Real-life networks 45
4.5   Scale-free networks 46

5. Self-organization and Symmetry   47
5.1   Growth of a crystal as an ordering process 47

5.2   Similar linkage patterns and symmetry 49
5.3   Symmetry as a secondary organizing principle 50
5.4   Symmetry and biology 52

6. The Different Types of Exact and Approximate Symmetry   59
6.1   Crystallographic symmetry 59
6.2   Space symmetry and time symmetry 60
6.3   Permutational and more general symmetries of graphs 60
6.4   Approximate symmetry of graphs 61
6.5   Symmetry in real-life networks 62
6.6   Structural *vs.* statistical equivalence and latent symmetry 69

7. Symmetry of Composite Systems   71
7.1   The Curie principle 71
7.2   The Curie-Shubnikov principle 73
7.3   Interplay between dissymmetrization and symmetrization 77
7.4   The Hermann theorem of crystal physics, and its applications 77
7.5   Hexply configurations for nanocomposites 79

8. Gauge Symmetry   81
8.1   Introduction 81
8.2   Gauge-symmetry groups 84
8.3   Noether's theorems 86

9. Phase Transitions and Broken Symmetry   93
9.1   Liberal meanings of the term 'phase transition' 93
9.2   Spontaneous breaking of symmetry 94
9.3   The Landau theory of phase transitions 95
9.4   Ferroic phase transitions and domain structure 97
9.5   Prototype symmetry 98
9.6   The symmetry compensation law 98
9.7   Continuous broken symmetries 99
9.8   Discrete broken symmetries 105
9.9   Broken symmetry and biology 105
9.10   The principle of local activity 108

**10. Particle Physics, Cosmology, and the Search for New
        Symmetries   111**
10.1   The Standard Model of Particle Physics 111
10.2   Beyond the Standard Model 121
10.3   Origin of our universe 124

11. Latent Symmetry, Potential Symmetry, and the Symmetry Composition Principle **129**
11.1 Latent symmetry and potential symmetry 129
11.2 The distinction between potential symmetry and latent symmetry 132
11.3 The fundamental theorem of symmetry 133
11.4 The symmetry composition principle 133
11.5 Placement symmetry 136
11.6 Latent symmetry and algorithmic information 137

12. Group-Theoretical Determination of Latent Symmetry **139**
12.1 Formal definition of latent symmetry 139
12.2 Litvin's partition theorem for latent symmetry 140
12.3 Latent symmetry and domain-average engineered ferroic materials 144
12.4 An example of how ignorance about latent symmetry can lead to errors 145
12.5 The role of placement symmetry in revealing latent symmetry 148
12.6 Concluding remarks 150

13. Symmetry of Complex Networks **151**
13.1 Latent symmetry in complex networks 151
13.2 Measures of symmetry of networks 154
13.3 Origins of symmetry in complex networks 156
13.4 The similar-linkage-pattern model for symmetry 157
13.5 The free-energy landscape for biological networks 158
13.6 Social networks and the meaning of cohesive energy 160

14. Afterword **163**

Bibliography **167**

Index **179**

Acknowledgements **187**

About the Author **189**

Foreword to the First Edition

Wherever we look we see a variety of patterns and shapes that show different types of symmetry. Much of this is obvious, such as for instance when we look at the pyramids of Egypt, or crystals in a museum. However, what is not so obvious is just what exactly is symmetry and why is it so prevalent? In this unique and intriguing book, Professor Vinod Wadhawan has set about answering these sorts of questions. He takes us on a journey from very basic descriptions, such as the growth of a crystal, on to more esoteric and complex notions, demonstrating that, in fact, symmetry is even more pervasive than we thought before. Some symmetries are far from obvious, as illustrated by the idea of latent symmetry. This is said to manifest itself when one combines two or more 'equal' objects or systems, each with its own symmetry description, and the resulting composite system exhibits *new* symmetry elements that were not expected from the original systems. For instance, two identical right-angled isosceles triangles can be joined together to form a square, that has an unanticipated four-fold rotational symmetry. The notion of latent symmetry is relatively new and deserves further consideration.

Not only do we have the symmetry exhibited by living organisms and physical objects, but also by ideas themselves. As such this book has a strong philosophical content that will enable the reader to gain much more insight into the phenomenon than is normally got from a typical university education. Wadhawan shows us how even the concept of randomness is intricately bound up with notions of symmetry. Even the idea of predictability is an example of symmetry in action! And then, having explained what symmetry is, emphasis is placed on what happens when symmetry is broken. In a sense, pure symmetry could even be described as rather boring, since it implies a lack of change or progress. Nonetheless, we still need to understand it. It is when symmetry is broken that fun things start to happen and new ideas, progress and phenomena are created. This book explains how this comes about and why symmetry-breaking is so important. The book is written with an eye to explaining the fundamental concepts of symmetry, rather than go into complex mathematical proofs and lemmas, which in any case can be found elsewhere for those who like those sorts of things. This means that Wadhawan is able instead to concentrate on the philosophical importance of understanding symmetry, and how it impacts on the world that we observe. Rather like the Second Law of Thermodynamics, symmetry is seen to play a vital role in what holds the universe together. You can see then that this book covers just about everything that we know about symmetry, and possibly that which we do not!

A.M. Glazer
Professor of Physics and Emeritus Fellow of Jesus College
University of Oxford
Author of *Space Groups for Solid State Scientists*

July 2011, Oxford

Preface to the First Edition

The symmetry of any composite system made up of equal or equivalent components depends on at least two factors: The inherent symmetry of each component, and the symmetry imposed on the system by the manner in which the components are arranged with respect to one another ('placement symmetry'). But if the composite system is found to have a *higher* symmetry than what can be accounted for by these two factors, then that extra, unexpected symmetry is what I call *latent symmetry*. It is as if this additional symmetry was lying latent or dormant in the equal or equivalent components, and became manifest only when the components came together to form the composite system. To accommodate such a possibility, I enunciate in this book a new *symmetry composition principle*. According to it: When the occurrence of a symmetry implies the coexistence of two or more equal or equivalent building blocks, the overall symmetry is either the product of the building-block-symmetry group and the placement-symmetry group, or there is an *additional* component which arises from the latent symmetry present in the building blocks.

The emergence of symmetry in thermodynamically open composite systems can be traced ultimately to the second law of thermodynamics, which is therefore the *primary* organizing principle. How this principle operates in various diverse systems is discussed in this book. It is argued that the same explanation holds, whether it is the symmetry of a crystal, or that of a complex social network.

Symmetry of complex networks is, in fact, another major theme of this book. That real-life networks should possess any symmetry at all may come as a surprise. But by now we should all be reconciled to the fact that there is something about symmetry which touches everything in our universe. The present edifice of science in general, and physics in particular, would be unthinkable without symmetry. There is a lot of symmetry even in biological systems.

We are surrounded not only by symmetry, but also broken symmetry. In fact, we see more of broken symmetry than intact symmetry. From the Big Bang onwards, as our universe cooled and expanded, a series of symmetry-breaking transitions occurred, leading eventually to the complexity of life we see today. This book is an attempt to explain, in an accessible language, the interplay between latent, manifest, and broken symmetry.

Bangalore   Vinod Wadhawan
August 2011

Preface to the 2014 Print

This print incorporates some minor corrections and other alterations.

Bangalore   Vinod Wadhawan
April 2014

Preface to the Second Edition

The book has been revised and updated substantially. In particular, gauge symmetry, which was discussed only briefly in the first edition, has been given the prominence it deserves. A new chapter has been added to deal with it in some detail.

Another new feature of this edition is the introduction of my notion of *potential symmetry*. It is similar to latent symmetry, but not identical to it. Latent symmetry is a kind of potential symmetry which becomes manifest symmetry when the conditions are just right. But potential symmetry is not always latent symmetry; in fact, it is only rarely so. Introduction of the notion of potential symmetry enables us to enunciate what I call *the fundamental theorem of symmetry*. It says that any spontaneously occurring symmetry of an object or system comprising of equal or equivalent subparts is nothing but a self-organized manifestation of the potential symmetry residing in its subparts.

Bangalore   Vinod Wadhawan
July 2018

1. Overview

Symmetry, as wide or as narrow as you may define its meaning, is one idea by which man through the ages has tried to comprehend and create order, beauty and perfection.

Hermann Weyl

We are surrounded by symmetry, and broken symmetry. Great progress has been made in describing, understanding, and exploiting the symmetry and broken symmetry we observe in Nature, but we seldom pause to wonder *why* the symmetry exhibited by an object or phenomenon is what it is, and not some other symmetry. And why is it that, for example, organic molecules, with generally *no* symmetry of their own, spontaneously form crystals which usually have a high degree of symmetry?

Another question of fundamental importance discussed in this book is: To what extent is symmetry an *organizing principle* in natural phenomena?

Crystals are *simple systems*, in the sense that they exhibit a high degree of uniformity or sameness, in contrast to the situation in *complex systems*. Complex systems are characterized, by definition, by *large variability* (Bak 1996; Wadhawan 2010, 2017). This variability arises from the nonlinear interactions among the large number of constituents of a complex system. Understanding the symmetry and other properties of complex systems has been greatly aided by the approach whereby the system is regarded as a *network*. The constituents of the complex system are taken as the *nodes* of the network, and an interaction between any two nodes is represented by an *edge* connecting those two nodes. This approach to complexity has the advantage that the formidable power of modern network theory can be brought to bear on the attack on the difficult problem of complexity.

Real-life complex networks (CNs) have an additional feature not usually present in mathematically (or deterministically) constructed networks: The number of nodes in them usually grows as the network evolves with time. A remarkable thing discovered mainly in and after the first decade of the 21st century is that real-life CNs usually possess a high degree of permutation symmetry (MacArthur and Anderson 2006; MacArthur, Sánchez-García and Anderson 2007, 2008; Hua *et al*. 2008; Garlaschelli, Ruzzenenti and Basosi 2010).

Why should symmetry arise in CNs? Is symmetry important − even central − to the self-organization tendencies of a complex system? Further investigations will throw light on this question, but there already are signs of progress. For example, the work of Xiao *et al.* (2008a, b) has shown that *'similar linkage patterns'* at a local level are responsible for the rich symmetry exhibited by many real-life CNs. Similarly, MacArthur *et al.* (2007) put forward the idea of *'growth with preferential attachment'* for explaining the occurrence of symmetry in certain real-life CNs. As I argue later in this book, growth with preferential attachment, as also similar linkage patterns, are also the reasons why crystals develop symmetry. There is clearly a need to investigate this aspect of symmetry more thoroughly.

In this book I also build up a case for my assertion that the 'why' part of the symmetry observed in Nature may sometimes have to do with what I called *latent symmetry* in an earlier book (Wadhawan 2000). Symmetry often implies the existence of two or more equal or equivalent subparts. And any overall manifest symmetry is indicative of three underlying factors:

(i) the symmetry manifestly present in the equal constituents of the object or phenomenon.

(ii) The way the equal constituents are located or placed with respect to one another (I call it *'placement symmetry'*); and

(iii) a possible symmetry that may lie *latent* in the equal constituents of the object or phenomenon, and becomes manifest only when the equal constituents are brought together in accordance with the prevailing placement symmetry.

The latent-symmetry part has been given a miss in the enormous literature on symmetry, and is not covered in any of the existing text books on symmetry.

Life forms exhibit a variety of symmetry (Fig. 1.1), and it can be instructive trying to understand the reasons for the observed symmetry (Ball 1999a, b).

Of course, we do sometimes have answers to questions regarding the *mechanism* involved or the *origin* of certain symmetries. For example, human beings and most other mammals have a vertical mirror plane of symmetry, or *'bilateral symmetry'*: The left part is roughly the same as the mirror image of the right part (Fig. 1.2(a)). This is easy to explain. There is

Fig. 1.1 Symmetry of some life forms. Image credit:
http://commons.wikimedia.org/wiki/File:SymmetryOfLifeFormsOnEarth.jpg

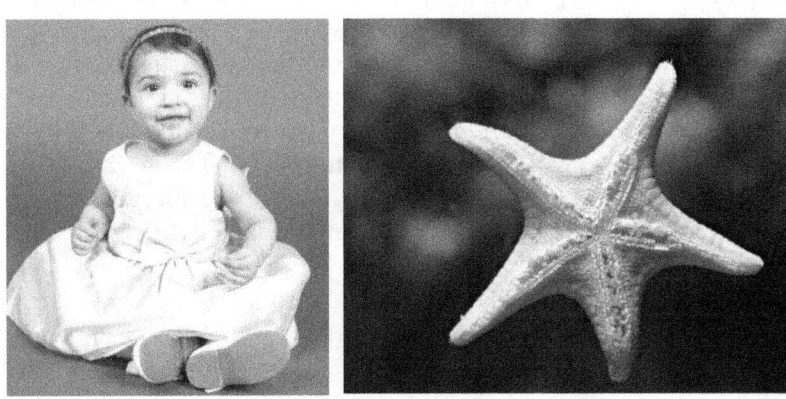

Fig. 1.2 (a) Humans and most other mammals have mirror-symmetric body shapes. (b) The ventral face of a sea star, showing tube feet. For creatures living in water, there are more options of locomotion modes than on land. Moreover, the downward pull of gravity is neutralized to a large extent by the upward force of buoyancy. The net reduced gravity does not demand very strictly the evolution of bilateral symmetry of body shape generally seen on land. Image credit:
http://commons.wikimedia.org/wiki/File:Starfish.jpg

the force of gravity, which defines a vertical direction. And there is the 'forward' direction which must be distinct from the 'backward' direction; otherwise the creatures would have an evolutionary disadvantage: A mammal which has forward-backward symmetry may not be able to decide quickly which way to run for escaping from danger. The forward horizontal direction and the gravitational vertical direction define a vertical plane, and we have evolved to have this plane of mirror symmetry in our body shapes.

The gravity argument gets support from the observed shapes of creatures that live in water. Many of them do have bilateral symmetry, but the weakened net downward gravitational force (because of the upward direction of buoyancy) increases the chances of evolution of shapes which have either no symmetry, or have uncommon symmetries not seen on land (Fig. 1.2(b)). An additional factor responsible for this is that the organisms can also roll over in water, apart from the more common forward direction of motion.

But here is an example of a seldom-addressed problem: Consider two types of organic molecules, neither of which has any intrinsic symmetry of its own. It can happen that one of them forms crystals which have a directional or point-group symmetry not present in the molecule. Its crystals may have space-group symmetry, say, P2. And the other type may crystallize with symmetry, say, P1, meaning that there is no directional or point-group symmetry present in the crystals (see, for example, Burns and Glazer (1978) for an introduction to the space-group symmetry of crystals). Naturally, the first molecule has some *potential* symmetry which lies dormant till a large number of molecules get a chance to come together and form a crystal. Such dormant symmetry needs to be investigated more, and I present my views on the inherent symmetry of things and phenomena in this book.

It is now widely believed that symmetry is fundamental to all scientific research at a very basic level (Rosen 2008). This becomes clear when one goes deep into the meaning of symmetry. I provide a glimpse of this in Chapter 2.

Group theory is the standard mathematical tool for describing symmetry, so I outline its basics in Chapter 3. And basic network theory is introduced in Chapter 4.

In Chapter 5 I go into some details of how a crystal grows from an aqueous phase, with the purpose of arguing why we can regard symmetry as a *secondary* organizing principle, the *primary* organizing principle being the second law of thermodynamics for open systems. The occurrence of symmetry in biological systems is also discussed in this chapter.

Chapter 6 presents a survey of the different *types* of symmetry. Real-life systems have an element of randomness, so it is necessary to introduce the idea of *approximate symmetry* for them. Both exact and approximate symmetries are reviewed in this chapter.

In Chapter 7 I deal with the symmetry of composite systems. The most fundamental principle for dealing with them is the Curie principle of summation of 'dissymmetries', and its generalization by Shubnikov. I had earlier given the name '*the Curie-Shubnikov principle*' to this generalized version of the Curie principle (Wadhawan 2000).

The term 'dissymmetrization' was used by Shubnikov and Koptsik (1974) for describing the lowering of symmetry that occurs when two unequal or different symmetries are superimposed. And 'symmetrization' meant a possibly higher symmetry than that envisaged in the original Curie principle, this increase occurring *only* when 'equal' or 'equivalent' objects are superimposed in some special ways. In Chapter 7 I also discuss the fine interplay between dissymmetrization and symmetrization we observe in Nature.

Chapter 8 is devoted to gauge symmetry. In common parlance, we implicitly assume that all our statements about symmetry are regarding *global* symmetry. By contrast, gauge symmetry is a *local* symmetry. But it is so fundamental to the understanding of the laws of physics that we may say that global symmetries are special cases of the relevant gauge symmetries, and not the other way around.

Broken symmetry and phase transitions are discussed in Chapter 9. An order parameter emerges at a phase transition in a crystal, and it has a certain symmetry of its own. The intersection of this symmetry with the original symmetry of the crystal breaks the symmetry of the crystal at the phase transition, in accordance with the Curie principle.

Whenever a symmetry operation is lost across a phase transition, it leaves its signature in the form of distinct domain types in the crystal. There is a kind of *symmetry-conservation principle* in operation here, because the lost symmetry operation is precisely the one that can map one domain type to another, give or take a little adjustment needed sometimes for matching the crystal structures at the domain boundaries.

Dynamical evolution and broken symmetry go together when we are dealing with open systems that are far from equilibrium. This is true of biological evolution also, a striking example being the asymmetry between the left and the right side of the human brain. Chapter 9 includes a brief discussion of such asymmetry also.

Broken-symmetry considerations form the backbone of particle physics. I explain in Chapter 9 how looking for (and finding) new broken symmetries leads to progress in unravelling the deepest secrets of Nature.

Chapter 10 is on particle physics, cosmology, and our search for new symmetries. The Standard Model of particle physics is described, particularly the role played by gauge symmetry in the development of this model. In modern cosmology also, gauge symmetry occupies a position of prominence. Even our search for new symmetries is influenced to a large extent by gauge-symmetry considerations.

My initial idea of latent symmetry (Wadhawan 2000) got a good group-theoretical treatment at the hands of Dan Litvin, who proved an important theorem in this context (Litvin and Wadhawan 2001, 2002). Chapters 11 and 12 give some details of this work. We also applied the latent-symmetry approach to the case of 'domain-average engineered' ferroic materials (Litvin, Wadhawan and Hatch 2003).

My attempts at treating manifest symmetry of geometrical systems as an outcome of the underlying placement symmetry and the inherent latent or potential symmetry of the equal subparts (the 'building blocks') have led me to formulate what I call *the symmetry composition principle*. According to it:

When the occurrence of a symmetry implies the coexistence of two or more equal or equivalent building blocks, the overall symmetry group is either the product of the symmetry group of the building block and the placement-symmetry group, or there is an <u>additional</u> symmetry which can arise from any latent symmetry present in the building blocks. [I have used the word 'group' in the above statement, in the group-theoretic sense. This should not be too restrictive. Admittedly, a more general treatment of the subject would also take note of the possibility wherein the symmetry operations do not always constitute a group.] The reasoning behind this formulation is outlined in Chapter 11. Fig. 1.3 is a graphic representation of this principle.

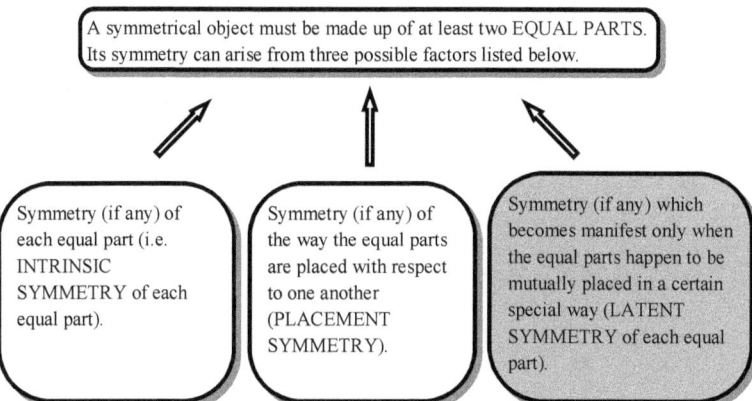

A symmetrical object must be made up of at least two EQUAL PARTS. Its symmetry can arise from three possible factors listed below.

Symmetry (if any) of each equal part (i.e. INTRINSIC SYMMETRY of each equal part).

Symmetry (if any) of the way the equal parts are placed with respect to one another (PLACEMENT SYMMETRY).

Symmetry (if any) which becomes manifest only when the equal parts happen to be mutually placed in a certain special way (LATENT SYMMETRY of each equal part).

Fig. 1.3 The symmetry composition principle. A symmetrical object often has two or more equal (or equivalent) parts. Its symmetry can arise from three factors: (i) each equal part may have an *intrinsic symmetry* of its own; (ii) the equal parts have a mutual disposition which may have some additional symmetry (called the *placement symmetry*); (iii) there may be *latent symmetry* in the equal parts, which becomes manifest in the overall object only for certain special placement symmetries of the equal parts. The placement symmetry plays a role both in the manifestation of the intrinsic symmetry of the equal parts, as well as in the manifestation of their latent symmetry.

Symmetry considerations have repeatedly led to leaps of understanding in not only mathematics and the various hard sciences, but also in other domains of human knowledge. An example, as mentioned above, is the rather recent identification of symmetry in many real-life complex networks. I discuss this in Chapter 13. It turns out that, once again, it is the inherent tendency towards *internal-energy minimization* or *cost minimization* or *effort minimization* that manifests itself as symmetry, as the network evolves in complexity.

The conventional approach to symmetry has been usually a *top-down* approach. We often speak of symmetries and broken symmetries. The phrase 'broken symmetry' is well known, and yet there is no equally well-known or standard phrase for the opposite of broken symmetry. '*Emergent symmetry*' seems to be a good opposite of broken symmetry (Ananthaswamy 2010). For understanding emergent symmetry, one needs to take a *bottom-up* approach. This book emphasizes such bottom-up processes responsible for the emergence of the observed symmetries.

2. Symmetry Fundamentals

Fundamental symmetry principles dictate the basic laws of physics, control the structure of matter, and define the fundamental forces of Nature.

Lederman and Hill, *Symmetry in Physics*

Einstein's great advance in 1905 was to put symmetry first, to regard the symmetry principle as the primary feature of nature that constrains the allowable dynamical laws. Thus the transformation properties of the electromagnetic field were not to be derived from Maxwell's equations, as Lorentz did, but rather were consequences of relativistic invariance, and indeed largely dictate the form of Maxwell's equations. This is a profound change of attitude.

David Gross (1996)

2.1 Definition of symmetry

We speak of symmetry in any situation in which we can rearrange things (positions in space, values of quantum fields, etc.) and still get the same answer for any question we may ask about the physics of rearranged system.

Consider Fig. 2.1(a). It is a scalene triangle in two dimensions. There is nothing symmetrical about it.

Next consider Fig. 2.1(b). It has some symmetry because we can identify two equal subparts (I and II) in it, and a reflection across the vertical line maps I to II, and II to I. The vertical line is therefore a *symmetry element* for this object, and a reflection across this line is a *symmetry operation*.

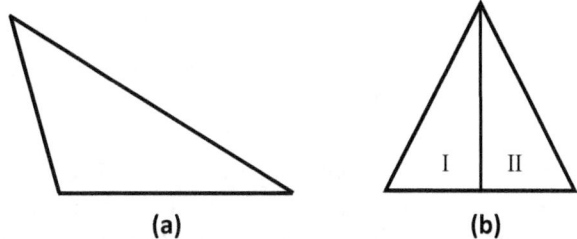

(a)   **(b)**

Fig. 2.1 (a) A scalene triangle in two dimensions. (b) An isosceles triangle in two dimensions.

Fig. 2.1(b) is an example of *geometrical symmetry*. The object is *embedded* in 2-dimensional space, and the symmetry operation is a coordinate transformation. But we can generalize and arrive at a definition of a symmetric system as follows:

A symmetric system has two characteristics (Rosen 2008):

1. There is a possibility of changing ('transforming') some parameter(s) relevant to the system.

2. The system remains the same or equivalent (or 'invariant') under this change of parameters.

In other words, *symmetry is invariance under a possible change*. In the example of Fig. 2.1(b) the possibility exists for performing a transformation, namely a mirror-reflection operation. And symmetry is present because the object is invariant (looks the same) after this operation has been performed.

In Fig. 2.1(b) a special aspect of symmetry is discernible, which may not be always present in other examples of symmetry: It is the presence of *equal (or equivalent) subparts*, such that the symmetry operations map one such subpart to another. To illustrate this point, let us consider a different example of a symmetrical system, namely the network shown in Fig. 2.2.

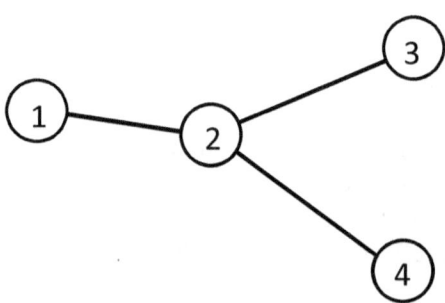

Fig. 2.2 A symmetrical network of four nodes (1, 2, 3, 4). It is symmetrical under the exchange of nodes 3 and 4.

There are four nodes (1, 2, 3, 4), with edges as shown. The topology remains invariant if the nodes 3 and 4 are interchanged: The number of edges remains the same, as also the connections around each node. This is an example of *permutation symmetry*. The topology of the set of nodes (1, 2, 3, 4) is symmetry-related to the topology of the set (1, 2, 4, 3).

2.2 Analogy and classification are symmetry

> *Indeed, science rests firmly on the triple foundation of reproducibility, predictability, and reduction, all of which are symmetries, with additional support from analogy and objectivity, which are symmetries too. So it is not much of an exaggeration to claim that science is symmetry. Or perhaps in somewhat more detail, science is our view of nature through symmetry spectacles.*
>
> Joe Rosen, *Symmetry Rules*

Analogy means invariance of the validity of a relation to changes of the elements involved in the relation. There is thus the possibility for change or transformation, and there is invariance under some transformations. Therefore, analogy is symmetry.

Kepler's work, leading to his three laws of planetary motion, is a good example of the power of using analogy in science. From the astronomical data available to him he concluded that the planetary data were analogous in a certain way which distinguished them from the data about other celestial bodies. Thus there were two distinct classes. For the first class he could formulate the three laws of motion which now go by his name. Each law coded the analogy observed in the behaviour as one substituted a planet in the solar system with another. This was a case of invariance (of the characteristics of motion) under the exchange of one planet by another.

What is more, analogy implies classification, and classification implies analogy (Rosen 2008). Thus, classification is also symmetry.

In chemistry the classification of elements according to the periodic table is another example of analogy being symmetry, and classification being symmetry.

2.3 Reduction is symmetry

Strictly speaking, it is impossible to identify any system which is completely isolated from other systems. Everything interacts with everything else, and thus *complexity* is the norm rather than the exception. This holistic way of looking at things, though valid and important, cannot always take us far in science because the problem becomes impossible even to formulate, let alone the question of solving it.

Complexity science attempts to get over this problem by adopting a variety of novel approaches (Wadhawan 2017/2018). It turns out that there is a whole spectrum of levels of complexity. At one end of the spectrum are the so-called simple or simplifiable systems for which it is a good approximation to adopt a *reductionistic* approach: Reductionism assumes that it is possible to understand any system in Nature as a sum of its parts: One can *reduce* a system to its ultimate subsystems, analyse and understand the subsystems *separately*, and then put together ('synthesize') the already understood subsystems to understand the whole system. This is clearly asking for too much, but it seems to work quite well for a whole class of systems. We call them *simple, or simplifiable, systems* (in contrast to complex systems).

What is done in the reductionistic approach is to define *quasi-isolated systems*. Quite often a judicious choice can be made of a subsystem that interacts only weakly with other systems or subsystems. Alternatively, it may be possible to account for the surroundings in an idealistic way, exemplified by the notion of a 'heat bath'. The heat bath is assumed to be very large compared to the subsystem in contact with it, so that any heat exchanged by the subsystem with the heat bath has negligible effect on the temperature of the heat bath. It is also assumed that the fact of making measurements on the system does not affect it in a significant manner.

If such a reduction is possible, it implies the presence of 'lawful' behaviour at the quasi-isolated system level, and therefore symmetry: The behaviour of the quasi-isolated system is invariant to whatever happens to the rest of Nature, and that means symmetry.

The assumption that it is possible to identify quasi-isolated subsystems is only one of the three ways of carrying out the reduction, and each implies symmetry of a certain kind (Rosen 2008). Let us take a look at all three.

Reduction to the observer and the observed

This is a major reduction indeed. The observer (*Homo sapiens*) is highly complex, and hardly understood. The 'observed' may be other humans, and this can again mean complications or poor understanding. But when we are observing macroscopic inanimate matter, this is usually a good approximation. An example is the explanation given by Newton of Kepler's laws. It was assumed by Newton that our observations have no effect on the motion of the planets. And, as stated above, Kepler's laws, or for that matter

Newton's laws (including the law of gravitation), are manifestations of symmetry, because the laws are invariant to where we study the system, or when we study the system.

But at the submicroscopic level, where quantum-mechanical effects cannot be ignored, the act of observation cannot be separated from what is being observed.

Reduction to quasi-isolated system and environment

Reduction to the observer and the observed is the primary or the most coarse level of reduction. 'The observed' may comprise the whole of Nature, which is still too large a subsystem. The division of Nature into a quasi-isolated system and the environment is the next level of reduction.

Those aspects of quasi-isolated systems that are invariant to changes in the environment comprise the symmetry of such systems. There may be additional symmetries within the systems.

Reduction to initial state and evolution

It is for quasi-isolated systems that order and law are found. Therefore, for such systems reduction into initial state and evolution of the initial state makes sense. This is a reduction along the time axis, unlike the above two which are reductions in space. The state of a quasi-isolated system at any instant of time is the initial state that changes into the next state at the next instant of time. And this latter state is the initial state for the state at the next instant of time, and so on. This is how an initial state evolves. And the law of this evolution is a manifestation of symmetry as a function of time: No matter what initial state is chosen the evolutionary behaviour is invariant to this choice.

In the case of our solar system, this kind of reduction was a factor responsible for the enunciation of Newton's three laws of motion, as also the law of gravitation.

2.4 Reproducibility is symmetry

In science we attempt to understand objectively the reproducible and predictable aspects of Nature. Reproducibility in science means that experiments can be replicated in the same laboratory at different times, as also in different laboratories at any time. *Reductionistic science rests on the double foundation of reproducibility and predictability.*

When the 'same' experiment is repeated at different times or places, several things about it are not identical. No two experiments are identical. But we can usually single out those factors that are relevant to the question of reproducibility. For example, the time of the day is one such parameter. Suppose we perform the experiment and obtain some result. We let an hour pass and repeat the experiment to obtain a result. If the two results are the same, after we have added one hour to the state of the experiment performed, and to the result obtained, we say that the experiment is reproducible.

Clearly this is also a manifestation of symmetry. We introduced a change or transformation (time delay), and the result is invariant to this transformation.

Reproducibility also implies analogy. The analogy here is that the new experiment is to the new result what the old experiment was to the old result. And, as already discussed, analogy is symmetry.

2.5 Predictability is symmetry

Since natural phenomena are governed by laws, we can conduct experiments and discover these laws. Predictability means that we can use the laws and predict the results of similar experiments not yet conducted.

Kepler's laws are a good example. When new planets were discovered after Kepler had formulated his laws, predictions could be made correctly about their motion around the Sun, in analogy with the behaviour observed and codified for the earlier planets.

This example also demonstrates that predictability implies analogy.

Predictability is symmetry: In the example of planetary motion, we can substitute data for another planet, and still find that the laws of motion are invariant to this substitution.

We can put all these considerations together, and say that because of reproducibility, predictability, and reduction, and aided by analogy, symmetry lies at the foundation of science. It may even be that science *is* symmetry.

2.6 The symmetry principle

The symmetry principle is often taken as first enunciated by Curie (1894). The originally formulated principle can be stated in three connected parts (see Koptsik 1983; Chiba and Nagahama 2001):

1. When several phenomena of different origin are superimposed in one and the same system, their *dissymmetries* are summed. There only remain the symmetry elements common to each phenomenon taken separately.

2. When certain causes lead to certain effects, the symmetry elements of the causes should be observed in these effects.

3. The statement contrary to these two conclusions is wrong, at least in practice; that is, the effects may be more symmetrical than the causes.

The first of these three statements is particularly relevant when we are interested in the symmetry of a *composite system*, and I shall make use of it directly in Chapter 7.

These statements have been much debated, and to the extent that they are valid, they embody the symmetry aspect of the *causality principle*. During the evolution of a process, the state of the system at any instant can be taken as the *cause*, and the state at a later instant the *effect*. According to the symmetry principle, the symmetry of the effect cannot be less than the symmetry of the cause. From this, one can enunciate a *symmetry evolution principle* (Rosen 1995):

The degree of symmetry of the state of a quasi-isolated system cannot decrease during evolution, but either remains constant or increases.

Curie's symmetry principle can also be stated as follows (Rosen 1995):

The effect is at least as symmetric as the cause.

Since the effect cannot be of a lower symmetry than the cause, Rosen (1995) turned the statement around to define a *lower bound* on the symmetry of the effect. He called this the *minimalistic use of the symmetry principle*.

Also, the *cause* cannot be of a higher symmetry than the effect. This defines an upper bound on the symmetry of the cause. This is the *maximalistic use of the symmetry principle*.

2.7 Thermodynamics and the symmetry principle

Why should the effect be at least as symmetric as the cause? Why is it that, during the dynamical evolution of an isolated system, symmetry cannot decrease? The answer lies in the second law of thermodynamics. The entropy of such a system cannot decrease, and therefore symmetry and entropy both increase as the system evolves with time.

Let us consider the example of ice and liquid water to understand the relationship between symmetry and disorder (or entropy).

Which is more symmetric, ice or water? Water is more symmetric. Ice is a crystal, so its invariance under rotations is confined to only certain specific (crystallographically self-consistent) rotations along certain specific directions. Thus, crystals are anisotropic objects, in general. By contrast, water looks the same from all angles. It is invariant under any and every rotation. Water is *disordered*. By contrast, an ice crystal is an *ordered* object.

A disordered system is more symmetric compared to its ordered version, if any.

And entropy is a measure of disorder. The entropy or the degree of disorder of an isolated system cannot decrease with time. Therefore the symmetry of an isolated system cannot decrease with time. It can either increase or stay the same. This is what the symmetry principle also says.

Nonequilibrium systems always have gradients of some kind or another. Natural phenomena are governed by the tendency to annul these gradients, as demanded by the second law of thermodynamics. Consider osmosis as an example. Suppose there is a membrane, with salty water on one side, and pure water on the other. The second law tells us that the difference in salt concentration will be reduced as time passes. On the whole, clean water will flow to the salty side and dilute it. This will happen spontaneously, not requiring the expenditure of any energy, as the system tries to reach a state of equilibrium (zero gradients).

But *reverse* osmosis requires that energy be spent. This is how seawater desalination is done. We have to spend energy to force salt to go to the more salty side across the membrane, to make seawater potable.

The forward and reverse osmosis processes are mirror-image processes, which reflect the time-symmetry or reversibility of the kinetics of motion at the microscopic level. A *reciprocity relation* connects the two processes, as demonstrated in the Nobel-Prize winning work of Lars Onsager in 1968.

The reciprocity relations are an example of how even nonequilibrium systems can be highly ordered, regular or symmetric.

2.8 Ugly symmetry

> *My theory of structural stability and process spontaneity is a new and a quantitative relation of the five concepts: higher symmetry, higher similarity, higher entropy, less information and less diversity; and they are all related to higher stability.*
>
> S. K. Lin (1999a)

Water has more symmetry than ice. Also, the entropy of water is more than that of ice because there is more disorder in the structure of water. Thus symmetry, entropy and disorder go together. If we associate order with beauty (as many people do), then disorder is ugly and symmetry is ugly! Ice is more 'beautiful' than water: Water is ugly by comparison because water is disordered and has higher symmetry. A symmetric structure is stable, but not necessarily beautiful. The term 'ugly symmetry' has indeed been used in serious scientific literature by Lin (1996a, 1999a, 2001). ' .. symmetry is in principle ugly, because it is related to entropy and information loss' (Lin 1996a).

Although logically correct from the point of view of a physicist or a chemist, the term 'ugly symmetry' may provoke howls of protest from certain quarters. I reproduce here the description of symmetry and beauty from the Wikipedia:

'**Symmetry** (from the Greek: "συμμετρεῖν" = to measure together), generally conveys two primary meanings. The first is an imprecise sense of harmonious or aesthetically pleasing proportionality and balance; such that it reflects beauty or perfection. The second meaning is a precise and well-defined concept of balance or "patterned self-similarity" that can be demonstrated or proved according to the rules of a formal system: by geometry, through physics or otherwise.'

'**Beauty** is a characteristic of a person, animal, place, object, or idea that provides a perceptual experience of pleasure, meaning, or satisfaction. Beauty is studied as part of aesthetics, sociology, social psychology, and culture. An "ideal beauty" is an entity which is admired, or possesses features widely attributed to beauty in a particular culture, for perfection. The experience of "beauty" often involves the interpretation of some entity as being in balance and harmony with nature, which may lead to feelings of attraction and emotional well-being. Because this is a subjective experience, it is often said that "beauty is in the eye of the beholder." In its most profound sense, beauty may engender a salient experience of positive reflection about the meaning of one's own existence. A subject of beauty is anything that resonates with personal meaning.'

It is difficult to correlate beauty with degree of symmetry.

Fig. 2.3 shows a flower. It has 5-fold rotational symmetry, and it is beautiful. Rotational symmetry is very common in flowers, and 5-fold symmetry is particularly common (Endress 1999).

Fig. 2.3 A flower with 5-fold rotational symmetry.

Orchids tend to possess mirror symmetry or 'bilateral' symmetry (Fig. 2.4). Are they more beautiful or less beautiful than flowers with 5-fold rotational symmetry?

Fig. 2.4 Orchids.
Image credit: http://en.wikipedia.org/wiki/File:Haeckel_Orchidae.jpg

A symmetrical face is considered to be more beautiful compared to a markedly unsymmetrical face: Imagine a face with the nose pointing prominently to one side. But then imagine a perfectly symmetrical face with extra-large ears. Is it beautiful or ugly? It may look ugly to us because most of us do not have extra-large ears. But in a world inhabited by humans having what we consider extra-large ears, nobody would find anything amiss with such ears. Perception of beauty can be a highly subjective matter, and also a matter of conditioning.

Many of us find slightly imperfect symmetry very appealing. Painters often introduce some *broken symmetry* deliberately into their work.

Static and dynamic symmetry

Suppose we have cameras which can image atoms very well, and can do so at speeds much greater than the typical speeds of atoms in water and ice. A snapshot of ice will show a very regular arrangement of atoms on a crystalline lattice. [We can ignore the small displacements due to thermal vibrations.] This is *static symmetry*.

On the other hand, a snapshot of liquid water will be highly unsymmetrical. But if we take a large number of such photographs and superimpose them, we get something which exhibits complete rotational symmetry. This is *dynamic symmetry*.

Lin (1996a, 2001) has demonstrated that his 'ugly symmetry' conclusion holds for both static and dynamic symmetries.

3. Group-Theoretical Description of Symmetry

Why is group theory so effective in describing the physical world? The answer is that it codifies the basic axioms of the scientific enterprise. The logic of group theory is the logic of scientific inquiry. Group theory is the mathematical formulation of internal consistency in the description of things. We assume that the system being observed has an intrinsic character independent of the observer's perspective. It's there. It possesses an objective reality. On this assumption - that it's there - how the system is perceived under altered scrutiny must be a matter of logic. Its appearance follows the logic of <u>intrinsic sameness</u>; that just because we change our point of view in examining the system does not change the nature of the system. The codification of that logic is a matter of group theory. And its success in portraying the physical world is what validates the <u>intrinsic sameness</u> assumption.

Marvin Chester

http://www.physics.ucla.edu/~chester/PhysAsSymtry/3Pane.html

3.1 Discrete groups

A group is a *set* for which there exists a rule for combining or 'multiplying' any two members of the set, and the set satisfies the following four conditions:

1. The set has the *closure* property: The product of any two members (or elements) of the set is also a member of the set.

2. The set has the *associativity* property: If A, B, C are any three members of the set, then $(AB)C = A(BC)$.

3. The set includes the *identity element*, say E, such that for any member M of the set, $ME = EM = M$.

4. For every M in the set there exits an *inverse element* M^{-1} in the set such that $MM^{-1} = M^{-1}M = E$.

The number of elements in a group is called its *order*.

If A and B are any two elements of a group (other than the identity element), and if B is not the inverse of A, then $AB \neq BA$, in general. If $AB = BA$, then the group is said to be an *Abelian group*. Otherwise, non-Abelian.

Let us consider the symmetry of a square in three dimensions (Fig. 3.1a). The x- and the y-axes are as shown, and the z-axis is perpendicular to the plane of the paper.

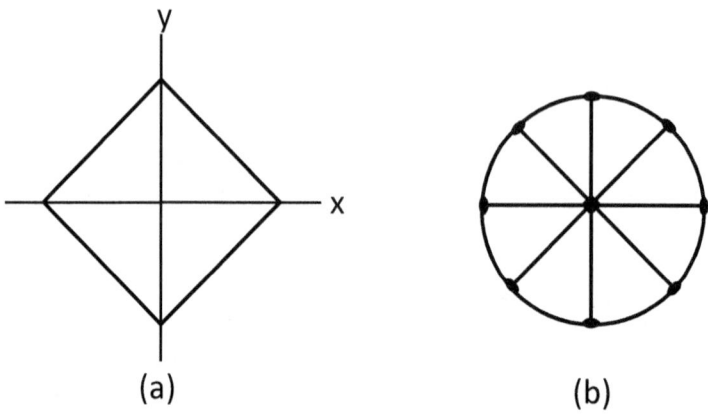

(a)   (b)

Fig. 3.1 A square in three dimensions (a), and its symmetry elements (b).

This object has a symmetry group which has 16 elements; i.e., the group is of order 16. There are 16 distinct coordinate transformations (symmetry operations) which map the square back onto itself. Fig. 3.1b shows the various symmetry elements for the square. In crystallography, the symbol used for this group is D_{4h} or 4/*mmm*.

Eight of the 16 operations are: $E, 2C_4, C_2, 2C_2', 2C_2''$. Here E is the identity operation in the group; it corresponds to performing no coordinate transformation at all. There for two 4-fold axes of symmetry along the z-axis, one performing a rotation of 90°, and the other performing a rotation of -90°. There are five 2-fold axes of symmetry: One each parallel to the z-axis, the x-axis, and the y-axis; one at an angle of 45° to the x-axis in the xy-plane; and one at an angle of -45° to the x-axis in the xy-plane.

The other 8 elements of the group can be obtained from the above 8 by multiplying each of them by the inversion operation i. An inversion operation takes every point (x, y, z) to a point $(-x, -y, -z)$, if the origin is at $(0, 0, 0)$. We get: $i, 2S_4, \sigma_h, 2\sigma_v, 2\sigma_d$. Here S_4 denotes a 4-fold *roto-*

inversion axis of symmetry along the z-axis: The symmetry operation is a composite operation, comprising of a 4-fold rotation *and* an inversion operation. σ_h is a 'horizontal' plane of mirror symmetry, normal to the vertical z-axis. Similarly, σ_v denotes a 'vertical' plane of symmetry; i.e., it is parallel to the z-axis. And σ_d is a vertical 'diagonal' plane of mirror symmetry, making an angle of $45°$ to the zx-plane.

It turns out that the first eight symmetry elements also constitute a group by themselves; this set of elements satisfies all the four requirements for defining a group. The symbol for this (smaller) group is C_{4v}, or $4mm$. We say that it is a *subgroup* of the larger group D_{4h}.

3.2 Coset decomposition of a group

It is possible to *decompose* the group D_{4h} in terms of its subgroup:

$$D_{4h} = C_{4v} + i(C_{4v}) \tag{3.1}$$

This is an example of *coset decomposition*. The elements of D_{4h} can be divided into two parts (cosets), and one part can be obtained from the other by multiplying by an element (i in the case chosen above) which is not a member of the other part.

In general, let G be a group of order g, and let $H\{h_1, h_2,...h_h\}$ be a nontrivial or 'proper' subgroup of G. There exists at least one element a of G which is not in H. Then the set ($ah_1, ah_2,...ah_h$) is a *left coset* of H by a.

Similarly ($h_1a, h_2a,...h_ha$) is a *right coset* of H by a. Since, in general, $AB \neq BA$ for a group, the left coset is not necessarily identical to the right coset. But the two cosets are not necessarily disjoint.

The element a is a *coset representative*. Its choice is not unique. Any member of the coset can be a coset representative.

Next let us consider the symmetry group of a rhombus (Fig. 3.2). The most noticeable missing element of symmetry now is the vertical 4-fold axis. The symmetry group has only 8 elements now, and is denoted by the symbol D_{2h} or *mmm*. Its elements are: $E, C_2, C_2', i, \sigma_h, \sigma_v, \sigma_v'$.

D_{2h} is a subgroup of D_{4h}. And the following is a valid coset decomposition:

$$D_{4h} = D_{2h} + C_2{}'(D_{2h})$$
(3.2)

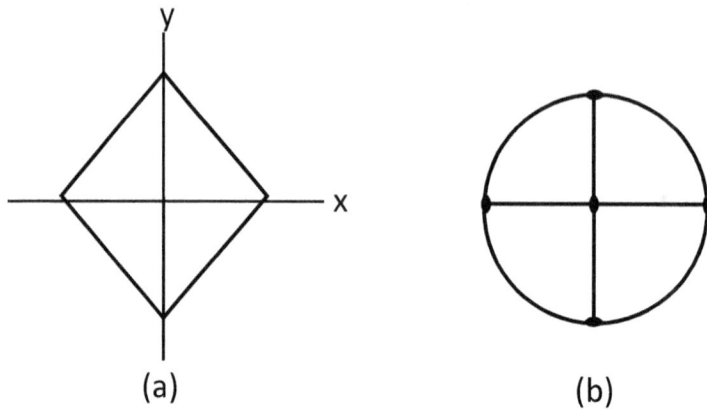

(a)   (b)

Fig. 3.2 A rhombus in three dimensions (a), and its symmetry elements (b).

When a subgroup H is such that for its supergroup G the left coset and the right coset are the same, it is called a *normal subgroup*. In other words, a normal subgroup is one which is invariant under conjugation by members of its supergroup: $gH = Hg$ for all g not in H.

A *simple group* is a nontrivial group, of which the only normal subgroups are the trivial group and the group itself.

Any non-simple group can be broken into two smaller groups, a normal subgroup and the *quotient group* (also called a *factor group*). The process can be repeated.

A quotient group or a factor group is that which is obtained by aggregating similar elements of a larger group using an equivalence relation that preserves the group structure. Formally, if H is a normal subgroup of G, then G/H (G over H) is the quotient group of G w.r.t. H, the members of the quotient group being the cosets of G w.r.t. H. Each coset is collectively an element of the group, and the multiplication of two such elements amounts to multiplying each member of one coset with each member of the other coset (counting repeated elements only once).

The general definition is as follows: If H is a normal subgroup of G, the set of all the distinct cosets of H in G, together with the coset multiplication rule explained above, is called the factor group or the quotient group of G with respect to H. It is denoted by K = G/H.

3.3 Lagrange theorem for subgroups

As seen above, any two distinct cosets of a group G are always disjoint. Moreover, in each coset the number of elements is equal to the order of the subgroup H. It follows that the g elements of the group G can be split into an *integral* number n of disjoint sets:

$$g = nh, \tag{3.3}$$

where h is the order of the subgroup H. This is a general result, and is known as the Lagrange theorem for subgroups. It says that:

The order of any subgroup H of G is a divisor of the order of the order of G.

The number n is called the *index* of H in G.

3.4 Symmetry group of a crystal

One can specify the structure of a crystal in terms of a *density function* $\rho(x, y, z)$, which defines the electron density at every point (x, y, z) in the crystal. There can be certain distance-preserving coordinate transformations (translations, rotations, reflections, inversions, or their self-consistent combinations) which map the density function back onto itself; i.e., they leave it invariant. They are therefore called *symmetry transformations*. The set of all such symmetry transformations constitutes a group, called the *symmetry group* of the crystal.

Nonmagnetic crystals can be grouped into 7 crystal systems, 14 Bravais lattices, 32 point groups, and 230 space groups. Additional symmetry operations have to be identified for dealing with magnetic crystals (see, e.g., Wadhawan 2000).

All crystals possess at least the translational symmetry, restricted to one of the 14 Bravais lattices. Translational symmetry means that if there is something, say an atom, at a position $\mathbf{r}$ in the crystal, the same atom must also be found at $\mathbf{r}+\mathbf{t}_n$, where

$$\mathbf{t}_n = n_1\mathbf{a} + n_2\mathbf{b} + n_3\mathbf{c} \tag{3.4}$$

Here the integers n_1, n_2, n_3 can take any value between $-\infty$ and $+\infty$. And $(\mathbf{a}, \mathbf{b}, \mathbf{c})$ are a triplet of noncoplanar vectors called the *basis vectors*. They define the three edges of a parallelepiped called the *unit cell*. If the unit cell (along with its content of atoms) is stacked repeatedly along the directions of the three basis vectors in a space-filling manner, one obtains the entire crystal. The unit cell is specified by the lengths (a, b, c) of its three edges and the angles (α, β, γ) between its edges; e.g., α is the angle between $\mathbf{b}$ and $\mathbf{c}$. In a rectangular unit cell, all three angles would be 90°.

The presence of translational symmetry restricts the directional symmetry of crystals to be from among the 32 point groups only.

The full symmetry group of a nonmagnetic crystal can only be from among the 230 space groups. To describe the space-group symmetry of a crystal, we have to identify its Bravais lattice, as also the symmetry elements involving rotations and reflections (including any 'screw' axes and 'glide' planes). The relative locations of the symmetry elements in the crystal lattice must also be identified.

A simple example of a crystallographic space group is the one denoted by the symbol P2. Here the symbol 2 denotes the directional symmetry, or the point-group symmetry. It means that in the crystal there is a direction (say the c-axis, or the z-axis) which is an axis of 2-fold symmetry: If we rotate the crystal by an angle of $2\pi/2$, or 180°, we get back the same crystal structure as before the application of this symmetry operation. Similarly, an n-fold symmetry axis implies invariance under a rotation of $2\pi/n$. The translational symmetry of all crystals limits n to have only the values 1, 2, 3, 4, or 6. Since there is only one 2-fold axis indicated in the symbol P2, it means that the crystal belongs to the *monoclinic crystal system* (from among the seven crystal systems). If the z-axis is chosen to lie along this 2-fold axis, the shape of the unit cell is such that $a \neq b \neq c$ and $\gamma = 90^0 \neq \alpha \neq \beta$.

Next, consider the space group P1. This symbol stands for the lowest symmetry conceivable for a crystal. There is no directional symmetry present. And the translational symmetry is such that $a \neq b \neq c; \alpha \neq \beta \neq \gamma$. This space group thus corresponds to the so-called *triclinic crystal system*.

3.5 Continuous groups

If the order of a group (i.e., the number of elements in it) is nondenumerably infinite, it is called a continuous group.

A familiar example of a continuous group is the set of all real numbers. It is called a continuous group of *order* 1, because only one number, say x, is sufficient to specify any real number in the range $[-\infty, \infty]$.

3.6 Permutation groups

A transformation defined on a *point field* is called a permutation. [A point field is a set of elements called *points*.]

A *mapping* from a point field A to a point field B is said to be defined if for every point p in A, a point p' is associated in B.

$p'=f(p)$ is called the *image* of p. Such a mapping also defines the *function f*.

A *permutation* is a transformation defined on a finite point field.

If a set of permutations satisfies the four conditions for defining a group, it is called a *permutation group*.

3.7 Special unitary groups

Let R be a matrix, and R^T its transpose. If the matrix product RR^T is the same as the product $R^T R$, and both are equal to a unit matrix I, then R is said to be an *orthogonal matrix*:

$$RR^T = R^T R = I \tag{3.5}$$

Since the matrices R and R^T have the same determinant ($\det R$), Eq. 3.5 gives $(\det R)^2 = 1$, or

$$\det R = \pm 1 \tag{3.6}$$

The group of all 3×3 orthogonal matrices is a continuous group called *the orthogonal group O(3)*. Orthogonal matrices represent length-preserving or orthogonal transformations in 3-dimensional real vector space. Therefore the group of such transformations (also denoted by $O(3)$) is isomorphic to the group of orthogonal matrices. $O(3)$ is thus *the rotation-inversion group* in three dimensions.

The matrices of the group $O(3)$ can be divided into two sets: those with $\det R = +1$ (i.e., *proper* rotations), and those with $\det R = -1$ (i.e., *improper* rotations). The first set constitutes a group, $SO(3)$, called the *special orthogonal group* in three dimensions.

The set (R) of all real numbers constitutes a *field*, as also does the set (C) of all complex numbers. The elements of a field are called *scalars*.

The set of all non-singular, square, distinct matrices of any order forms a group, with matrix multiplication as the law of composition for the elements of the group. Moreover, a square matrix $[T_{ij}]$ of order n provides a *representation* of an operator T in a basis (e_i) in a linear vector space of dimensionality n. Matrix sets are therefore used as representations of groups.

A *unitary matrix* U is an $n \times n$ complex matrix satisfying the condition

$$U^{\#}U = UU^{\#} = I_n \tag{3.7}$$

Here $U^{\#}$ is the conjugate transpose (or the Hermitian adjoint) of U, and I_n is the $n \times n$ unit matrix. This condition implies that the inverse of the matrix U is equal to its conjugate transpose:

$$U^{-1} = U^{\#} \tag{3.8}$$

The set of all $n \times n$ unitary matrices forms a group called the *unitary group*, $U(n)$.

The subset of all such matrices for which the determinant has the value $+1$ forms a group called *the special unitary group, SU(n)*.

When all elements of the matrices are real (rather than complex) numbers, the special unitary group $SU(n)$ is the same as *the special orthogonal group, SO(n)*.

3.8 Topological space, open sets

A topological space is a set of points, along with a set of 'neighbourhoods' for each point, with a specified set of axioms relating points and neighbourhoods. [A *neighbourhood of a point* is a set of points containing that point where one can move some amount in any direction away from that point without leaving the set.]

Other spaces such as manifolds and metric spaces are special cases of topological space, involving additional structures or constraints. [A *manifold* is a generalization of concept of a curved surface. It is a topological space modelled on Euclidean space locally, but may vary widely in global properties. Each manifold is equipped with a family of local coordinate systems related to each other by coordinate transformations belonging to a specified class.]

There are many alternative ways of defining a topological space, depending on the choice made regarding the set of axioms relating points and neighbourhoods. Here is a definition via what are called '*open sets*' (see below):

A topological space is an ordered pair (X, T), where X is a set and T is a collection of subsets of X, with T characterized by the following three axioms:

1. The null or empty set, and the full set X, both belong to T.
2. Any finite or infinite union of members of T belongs to T.
3. The intersection of any finite number of members of T belongs to T.

The elements of T are called open sets. And T itself is called a topology on X.

3.9 Morphisms, categories

A *morphism* (http://mathworld.wolfram.com/Morphism.html) is a mapping between two *objects*. [An object is a mathematical structure, e.g. a group, vector space, or smooth manifold in a 'category' (see below).]

A general morphism is called a *homomorphism*.

A morphism $f: Y \to X$ in a 'category' is a *monomorphism* if, for any two morphisms $u, v: Z \to Y$, $fu = fv$ implies that $u = v$. [*A category is an algebraic structure similar to a group, but without requiring inverse or closure properties.* For more details, see below.]

A morphism $f:Y \rightarrow X$ is an *epimorphism* if, for any two morphisms $u, v: X \rightarrow Z, uf = vf$ implies $u = v$.

A *bijection*, *bijective function*, *bijective morphism*, or *one-to-one correspondence* is a function between the elements of two sets, where each element of one set is paired with exactly one element of the other set, and each element of the other set is paired with exactly one element of the first set. It is also called an *isomorphism*. If there is an isomorphism between two objects, they are said to be *isomorphic*.

An isomorphism between an object and itself is called an *automorphism*.

A *category* consists of three things: (i) a collection of objects; (ii) for each pair of objects a collection of morphisms (sometimes call '*arrows*') from one to another, and (iii) a binary operation defined on compatible pairs of morphisms, called *composition*. The category must satisfy an identity axiom and an associative axiom, but without necessarily satisfying an inverse axiom and a closure axiom (http://mathworld.wolfram.com/Category.html).

A category consists of a class ob(C) of objects, and a class hom(C) of morphisms or arrows or maps between the objects. Each morphism f has a source object a and a target object b, where a and b are in ob(C).

A category C is called a *small category* if both ob(C) and hom(C) are actually sets and not proper classes. It is a *large category* otherwise. [A *class* is a collection of sets (or sometimes other mathematical objects) that can be unambiguously defined by a property that all its members share. A class that is not a set is called a *proper class*, and a class that is a set is sometimes called a *small class*. For instance, the class of all ordinal numbers, and the class of all sets, are proper classes in many formal systems.]

3.10 Semigroups, groupoids

A group is set with a binary operation that satisfies the four requirements of closure, associativity, identity element, and inverse element. Relaxation of one or more of these conditions leads to entities or objects such as semigroups and groupoids.

A *semigroup* is a set with a binary operation that satisfies only the associativity requirement. The other requirements of closure, identity element, and inverse element are not imposed.

Monoids are those objects that satisfy two of the four requirements, namely associativity and identity element.

We discuss *groupoids* next. A groupoid can be viewed as a group for which the binary operation is replaced by a 'partial function'. [A *partial function* from X to Y is a function $f : X' \rightarrow Y$ for some subset X' of X. Not every element of X is mapped to every element of Y.]

Another approach is to view it as a category in which every morphism is invertible. This type of category can be viewed as augmented with a unary operation called 'inverse' by analogy with group theory.

Groupoids have been defined in several ways (http://mathworld.wolfram.com/Groupoid.html).

One type of groupoid is an algebraic structure on a set with a binary operator, with closure being the only restriction on the operator. Associativity etc. are not imposed. An *associative* groupoid is nothing but a semigroup.

A second type of groupoid is, roughly, a category which is 'group-like' in the sense that every morphism or 'arrow' is an isomorphism, i.e., is invertible.

Before discussing the third type of definition of a groupoid, some more terminology must be introduced, namely the notion of base or basis.

In mathematics, a *base* (or *basis*) B for a topological space X with topology T is a collection of open sets in T such that every open set in T can be written as a union of elements of B. We say that the base *generates* the topology T. Bases are useful because many properties of topologies can be reduced to statements about a base generating that topology, and because many topologies are most easily defined in terms of a base which generates them.

A groupoid as an algebraic structure was first defined by Brandt in 1926, and is also known as a *virtual group*. His is the third type of definition of a groupoid, and can be shown to be equivalent to the second type above. According to it, a groupoid with base B is a set G with mappings α and β from G onto B, and a *partially* defined binary operation having certain axiomatic properties, namely associativity, inverse, and identity (what is missing here is closure, unlike the case of a group).

Any group is a groupoid with base a single point.

3.11 Lie groups

Lie groups lie at the intersection of algebra and geometry. Informally, a Lie group is a group of continuous symmetries, with some further provisos (see below). A circle has a continuous group of symmetries: you can rotate the circle by an arbitrarily small amount and it looks the same. This is in contrast to the hexagon, for example. If you rotate the hexagon by a small amount then it would look different. Only rotations that are multiples of one-sixth of a full turn are symmetries of a hexagon.

Lie groups capture and formalize the concept of continuous symmetries. They pertain to smoothly varying families of symmetries. Associated to any system which has a continuous group of symmetries is a Lie group.

Take the example of a sphere. It is invariant under any arbitrary rotation around any axis passing through its centre. The collection of all such coordinate transformations (the rotations) forms a Lie group, because the rotations can be made infinitesimally small, and any large rotation can be made up from a large number of such infinitesimal rotations.

For a symmetry group to be a Lie group, the equations which define its transformations must be *differentiable*. This is possible only for a locally smooth manifold, whereby the group operations are compatible with the smooth structure.

A *compact* Lie group is one for which the symmetries form a bounded set.

The basic building blocks of Lie groups are '*simple Lie groups*' (see below). For the classification of these groups one starts with the classification of the simple 'Lie algebras'. [A *Lie algebra* is a vector space together with a non-associative, alternating bilinear map, called the Lie bracket, satisfying the Jacobi identity. Lie algebras are closely related to Lie groups, which are groups that are also smooth manifolds, with the property that the group operations of multiplication and inversion are smooth maps. Any Lie group gives rise to a Lie algebra. Conversely, to any finite-dimensional Lie algebra over real or complex numbers, there is a corresponding connected Lie group unique up to 'covering'. This correspondence between Lie groups and Lie algebras allows one to study Lie groups in terms of Lie algebras.]

What is a simple Lie group? To answer that I should first introduce some concepts and definitions. Only a very brief description is given here. For more details, see, e.g., the Wikipedia.

Unit interval, path

In mathematics, the unit interval is the closed interval [0, 1], i.e., the set of all real numbers greater than or equal to 0 and less than or equal to 1. It is often denoted by I.

A 'path' in a topological space X is a continuous function f from the unit interval $I = [0, 1]$ to X ($f : I \rightarrow X$). It is a mapping. The *initial point* of the path is $f(0)$ and the *terminal point* is $f(1)$. One often speaks of a 'path from x to y' where x and y are the initial and terminal points of the path. A path is not just a subset of X which 'looks like' a curve; it also includes a 'parameterization'. For example, the maps $f(x) = x$ and $g(x) = x^2$ represent two different paths from 0 to 1 on the real line.

Pointed space

A pointed space is a topological space with a distinguished point, the '*base point*'. The distinguished point is one particular point, picked out from the space and given a name such as x_0, that remains unchanged during subsequent discussion, and is kept track of during all operations.

Fundamental group

In algebraic topology, the fundamental group is a mathematical group associated to any given pointed topological space that provides a way to determine when two paths, starting and ending at a fixed base point, can be continuously deformed into each other. It records information about the basic shape, or *holes*, of the topological space.

Simply connected space

A topological space is said to be simply connected (or 1-connected, or 1-simply connected) if it is path-connected and every path between two points can be continuously transformed, staying within the space, into any other such path while preserving the two endpoints in question. The fundamental group of a topological space is an indicator of a possible failure for the space to be simply connected: a path-connected topological space is simply connected if and only if its fundamental group is trivial.

Topological groups

A topological group is a group G together with a topology on G such that the binary operation of the group and the inverse function of the group are continuous functions w.r.t. the topology.

Every group can be made into a topological group by considering it with a discrete topology; such groups are called *discrete groups*. Example: real numbers with the usual topology form a topological group under addition.

Group objects

In category theory, group objects are certain generalizations of groups that are built on more complicated structures than sets. A typical example of a group object is a topological group whose underlying set is a topological space such that the group operations are continuous.

2-groups

A 2-group, or 2-dimensional higher group, is a certain combination of group and groupoid. The 2-groups are part of a larger hierarchy of *n*-groups. In some of the literature, 2-groups are also called *gr-categories* or *groupal groupoids*.

Much of the literature focuses on *strict 2-groups*. A strict 2-group is a strict monoidal category in which every morphism is invertible and every object has a strict inverse (so that xy and yx are actually equal to the unit object).

A strict 2-group is a group object in a category of categories; as such, they are also called *groupal categories*. Conversely, a strict 2-group is a category object in the category of groups; as such, they are also called *categorical groups*. They can also be identified with 'crossed modules', and are most often studied in that form. Thus, 2-groups in general can be seen as a weakening of crossed modules.

[In mathematics, a *module* is a fundamental algebraic structure used in abstract algebra. For example, a module over a ring is a generalization of the notion of vector space over a field, in which the corresponding scalars are the elements of an arbitrary given ring (with identity) and a multiplication is defined between elements of the ring and elements of the module. Thus, a module, like a vector space, is an additive Abelian group; a product is defined as between elements of the ring and elements of the module that is distributive over the addition operation of each parameter and is compatible with the ring multiplication. A *crossed module* consists of groups G and H, where G acts on H by automorphisms and a homomorphism of groups that is equivalent w.r.t. the conjugation action of G on itself, and also satisfies the so-called '*Peiffer identity*'.]

Every 2-group is equivalent to a strict 2-group, although this cannot be done coherently: it does not extend to 2-group homomorphisms.

If G is a strict 2-group, then the objects of G form a group, called the *underlying group* of G and written as G_0. This will not work for arbitrary 2-groups; however, if one identifies isomorphic objects, then the equivalence classes form a group, called the *fundamental group* of G and written as $\pi_1(G)$. (Note that even for a strict 2-group, the fundamental group will only be a quotient group of the underlying group.)

Simple groups

A simple group is a nontrivial group whose only normal subgroups are the trivial group and the group itself. A group that is not simple can be broken into two smaller groups, namely a normal subgroup and the quotient group, and the process can be repeated. If the group is finite, then eventually one arrives at uniquely determined simple groups by the so-called Jordan–Hölder theorem. The complete classification of finite simple groups, finished in 2008, is a major milestone in the history of mathematics.

Simply-connected groups

[Acknowledgement: Simply-connected group. *Encyclopaedia of Mathematics*. http://www.encyclopediaofmath.org/index.php?title=Simply-connected_group&oldid=42436].

A simply connected group is a topological group (in particular, a Lie group) for which the underlying topological space is simply connected.

The significance of simply-connected groups in the theory of Lie groups is explained by the following two theorems:

1) Every connected Lie group G is isomorphic to the quotient group of a certain simply-connected group (called *the universal covering of G*) by a discrete central subgroup isomorphic to the group $\pi 1(G)$.

2) Two simply-connected Lie groups are isomorphic if and only if their Lie algebras are isomorphic; furthermore, every homomorphism of the Lie algebra of a simply-connected group $G1$ into the Lie algebra of an arbitrary

Lie group $G2$ is the derivation of a (uniquely defined) homomorphism of $G1$ into $G2$.

The centre Z of a simply-connected semi-simple compact or complex Lie group G is finite. It is given in the following table for the various kinds of simple Lie groups.

G	A_n	B_n	C_n	D_{2n}	D_{2n+1}	E_6	E_7	E_8	F_4	G_2
Z	$\mathbf{Z}_{n+1}$	$\mathbf{Z}_2$	$\mathbf{Z}_2$	$\mathbf{Z}_2 \times \mathbf{Z}_2$	$\mathbf{Z}_4$	$\mathbf{Z}_3$	$\mathbf{Z}_2$	$\{e\}$	$\{e\}$	$\{e\}$

In the theory of algebraic groups, a simply-connected group is a connected algebraic group G not admitting any non-trivial 'isogeny' $\phi: \tilde{G} \sim G$, where $\tilde{G}$ is also a connected algebraic group. For semi-simple algebraic groups over the field of complex numbers this definition is equivalent to that given above.

Simple Lie groups

A simple Lie group is a connected non-Abelian Lie group G which does not have nontrivial connected normal subgroups.

A simple Lie algebra is a non-Abelian Lie algebra whose only 'ideals' are 0 and itself (or equivalently, a Lie algebra of dimension 2 or more, whose only ideals are 0 and itself).

Simple Lie groups are a class of Lie groups which play a role in Lie-group theory similar to that of simple groups in the theory of discrete groups. Essentially, simple Lie groups are connected Lie groups which cannot be decomposed as an extension of smaller connected Lie groups, and which are not commutative.

Together with the commutative Lie group of the real numbers, and that of the unit complex numbers, $U(1)$, simple Lie groups give the atomic 'blocks' that make up all (finite-dimensional) connected Lie groups via the operation of group extension. Many commonly encountered Lie groups are either simple or close to being simple: for example, the group $SL(n)$ of $n \times n$ matrices, with determinant equal to 1, is simple for all $n > 1$.

An equivalent definition of a simple Lie group follows from the Lie correspondence: a connected Lie group is simple if its Lie algebra is simple. An important technical point is that a simple Lie group may contain *discrete* normal subgroups, hence being a simple Lie group is different from being simple as an abstract group.

Simple Lie groups include many classical Lie groups, which provide a group-theoretic underpinning for spherical geometry, projective geometry and related geometries in the sense of Felix Klein's Erlangen programme (https://en.wikipedia.org/wiki/Erlangen_program). It emerged in the course of classification of simple Lie groups that there exist also several exceptional possibilities not corresponding to any familiar geometry. These *exceptional groups* account for many special examples and configurations in other branches of mathematics, as well as contemporary theoretical physics.

4. Network Theory

The notion of networks as a dominant organizing principle to explain how the world really works has attracted enormous interdisciplinary interest. Physicists are talking to mathematicians who are talking to sociologists and economists who are talking to physicists. In barely a decade, networks of researchers have sprung up to research networks.

http://news.cnet.com/2009-1069-978596.html#ixzz1QQBESzur

4.1 Mathematical networks

A *graph* is a collection of points (called nodes or vertices) together with a collection of lines (called edges or links) that connect certain pairs of points.

A *directed graph* is a graph in which the edges are *ordered pairs* of vertices; that is, every edge in a directed graph has a *direction*. A directed edge (often called an *arc*) runs only in one direction (e.g., a one-way road between two locations), and can be represented by indicating an arrow on it.

When the edges are unordered pairs of vertices, we speak of an *undirected graph* or a *simple graph*.

A *network* is a directed graph in which every edge is given a *label*. The term network is used in several different contexts, and can have different, context-dependent meanings. Electrical networks, social networks, communication networks, and neural networks are some examples. In the scientific literature, a distinction between graphs and networks is often not made.

A simple graph is an unordered set of vertices (nodes) and edges:

$$G = G(V, E) \qquad (4.1)$$

Here V is a set of vertices, and E is a set of edges with the proviso that

$$E \subseteq V \times V \qquad (4.2)$$

The number of nodes in a graph is called its *order*. That is, it is the cardinality of the set V, denoted by $|V|$, or card(V), or c(V).

The *size* of a graph is the number of edges in it, and is denoted by $|E|$, card(E), or c(E).

If the vertices of a graph G can be divided into two disjoint subsets G_1 and G_2 such that every edge in the graph G connects a vertex in G_1 only to a vertex in G_2, it is called a *bipartite graph*. Such a graph contains vertices of two distinct types, and edges join only unlike types of vertices.

If every vertex in G_1 is connected to every vertex in G_2, then G is a *complete bipartite graph*.

Consider two vertices u and v of G which are joined by an edge e. The edge $e = \{u, v\}$ is often represented as just uv (or u-v). We say that u is *adjacent with v*. And e is said to be *incident to u* and v.

Two edges incident to a common vertex are said to be *adjacent with each other*.

The *neighbourhood n(u)* of a vertex u is the set of vertices in the graph that are adjacent with u.

The number of vertices in the set $n(u)$ is called the *degree* of u. It is denoted by $d(u)$, and is the number of edges of the graph incident to u. A directed graph has both an *in-degree* and an *out-degree* for each vertex.

A graph is said to be a *finite graph* if it is of finite order.

A graph in which the degrees of all the nodes are finite is called a *locally finite graph*.

The *total degree* of a graph is the sum of the degrees of all its nodes.

Handshaking lemma: The total degree of a graph is equal to twice the number of edges in it.

Degree theorem: In any graph there is an even number of nodes with an odd degree.

If all vertices of a graph have the same degree r, then it is called an *r-regular graph*, or just a *regular graph*.

A graph H is a *subgraph* of a graph G if $|V(H)|$ is a subset of $|V(G)|$ and $E(H)$ is a subset of $E(G)$. G is called a *supergraph* of H.

If a graph G and its subgraph H have the same set of vertices, then H is called a *spanning subgraph* of G.

A *walk* W in a graph G from a vertex u to a distinct vertex v is a finite sequence of alternating vertices and edges ($W = v_0 e_1 v_1 ... v_k e_k v_{k+1}$, with $v_0 = u$ and $v_{k+1} = v$) such that each edge in the sequence is incident to each of the two vertices on either side of it in the sequence.

The *length* of the walk is the number k, i.e. the number of edges in it.

If all the edges in W are distinct, then it is called a *trail*.

A trail in which all the vertices are distinct is called a *path*.

A *geodesic path* is the shortest path between two vertices.

A walk, trail, or path is said to be *closed* if the initial and the final vertex are the same ($u = v$).

A closed path with $k \geq 3$ is called a *cycle*.

If there is a path between two vertices of a graph, then the vertices are said to be *connected*. If *each* pair of vertices in a graph is connected, then the whole graph is said to be a *connected graph*.

If only a subset of the vertices in a graph are connected, and if they are connected to a particular vertex, then this subset constitutes a *connected subgraph* (or a *cluster*).

A *disconnected graph* is one that is made of two or more isolated clusters.

The *distance* or *path length* between any two vertices u and v of a graph is the length of the shortest path between them.

The *diameter* of a graph or a network is the maximum distance (or length in terms of the number of edges) between any pair of nodes.

The average of the distances over all the pairs of nodes is the *average path length*. It is related to the size of the graph.

The *characteristic path length* of a graph is the median of the shortest paths from each node to every other node.

A *tree* is a connected graph that does not contain any cycles. It is readily argued that any connected graph has a spanning tree.

A *random graph or network* is one in which the edges are distributed randomly.

4.2 Clustering coefficient

Clustering or transitivity is an important feature of real-life networks. It is often found in such networks that if vertex A is connected to B, and B is connected to C, then there is a good chance that A is also connected to C. Your friend's friend is more likely to be your friend also. This transitivity means that, in the network topology, *triangles* may preponderate. [Triangles in a network are sets of three vertices, all connected to one another.]

Clustering in a network is an indication of deviation from random behaviour. For a random network, the 'clustering coefficient' is defined as simply the *average degree* $<k>$ of the network, which is also the probability p that any two nodes of the network are connected. All nodes in a random network have the same clustering coefficient:

$$C_{random} = p = <k> / N, \tag{4.3}$$

where N is the number of nodes in the network.

Clustering coefficients have been defined in more than one ways in the literature. Here is one such definition (Newman 2003):

$$C^{(1)} = \frac{3 \times \text{Number of triangles in the network}}{\text{Number of connected triples of vertices}} \tag{4.4}$$

Here 'connected triple' means a vertex with edges connecting it to an unordered pair of other vertices. Since each triangle contributes to three triples, the insertion of 3 in the above definition ensures that the clustering coefficient for the network lies in the range $0 \le C^{(1)} \le 1$. $C^{(1)}$ is the average probability that a friend's friend is a friend; it is the average probability that two vertices connected to a common vertex are connected to each other also.

Another widely used definition of the clustering coefficient is as follows (Watts and Strogatz 1998):

$$C_i = \frac{\text{Number of triangles connected to vertex } i}{\text{Number of triples centered on vertex } i} \qquad (4.5)$$

It is a *local* value of the clustering coefficient. For vertices with degree 0 or 1, we take $C_i = 0$. One can then define an average clustering coefficient for the network:

$$C^{(2)} = \frac{1}{N}\sum_i C_i \qquad (4.6)$$

Here N is the number of vertices in network.

4.3 Permutation symmetry in graphs and networks

A one-to-one mapping (or transformation) from the vertices of a graph G onto itself is a *permutation*.

Let $S(V)$ be the set of all the permutations possible for the set V of vertices comprising the graph G. In the set $S(V)$ there can be some permutations which preserve the *adjacency* of each vertex of the set V. These permutations are indicative of symmetry and are called *automorphisms*, acting on the set V of vertices of the graph G. The set of all automorphisms of the graph G is the *automorphism group*, denoted by Aut(G).

A network is said to possess *symmetry* if its underlying graph G has at least one automorphism which is not an identity or trivial permutation. If the identity permutation is the only automorphism of G, then the graph and the network is *asymmetric*.

A graph is said to be *vertex-transitive* or *node-transitive* if there is an automorphism between every pair of its vertices. In other words, graphs the automorphism group of which acts transitively on the set of vertices are vertex-transitive graphs.

A simple graph is said to be *edge-transitive* or *link-transitive* if there is an automorphism for all pairs of edges in it.

A simple graph is said to be *symmetric* if it is both vertex-transitive and edge-transitive.

A simple graph which is edge-transitive but not vertex-transitive is called a *semi-symmetric graph*.

A semi-symmetric graph is necessarily a bipartite graph.

A *Cayley graph* is a graph the automorphism group of which contains a subgroup that acts not just transitively, but *regularly*, on the vertices of the graph. This means that for any two vertices u and v of the graph, there is a *unique* automorphism in the subgroup that takes u to v (Alspach, Dobson and Morris 2008). Let H be this regular subgroup. Then the graph is said to be a Cayley graph on the group H.

One can define successive permutations or *products* of permutations. A product of two automorphism permutations f and g (written as fg) is another permutation h ($h = fg$). h is a mapping which takes the vertex set V onto itself, and this mapping has the same effect on a vertex x as if we first applied f on x (getting x^f), and then applied g on x^f (getting $(x^f)^g$). Thus,

$$x^h = (x^f)^g . \tag{4.7}$$

The set of automorphisms under the product of permutations forms a group, the *automorphism group* (Godsil and Royle 2001; Garrido 2011).

It can be proved that every group is the automorphism group of some graph. However, as observed by G. Pólya, not every group is the automorphism group of a tree.

A *partition* of a set V is a set of disjoint nonempty subsets of V whose union is V. These disjoint nonempty subsets are also called *cells*.

Given the automorphism group Aut(G) for the vertex set V, we can create a partition

$$P = \{V_1, V_2, ... V_k\} \tag{4.8}$$

such that a vertex x is equivalent to a vertex y if and only if Aut(G) contains a mapping g such that $x^g = y$. Such a partition is called an *automorphism partition*. And each cell of this partition is called an *orbit* of Aut(G).

A *trivial orbit* contains only a single vertex. Otherwise it is a nontrivial orbit.

Fig. 4.1 illustrates some of these ideas. There are 7 vertices in this graph, so the possible number of permutations is 7!, or 5040. Clearly, the adjacency of vertex 3 is different from that of 4. Therefore, a permutation that interchanges 3 and 4, and does nothing else, is not a symmetry transformation. So it is not an automorphism of this graph. A permutation that *is* an example of an automorphism is that in which only vertices 1 and 2 are interchanged. Thus, vertices 1 and 2 belong to the same orbit. Similarly, the vertices 5, 6 and 7 form a subset of the graph in which any interchange is an automorphism, so these three vertices belong to another orbit. We can find all the orbits, and write the automorphism partition as

$$P = \{\{1,2\},\{3\},\{4\},\{5,6,7\}\} \tag{4.9}$$

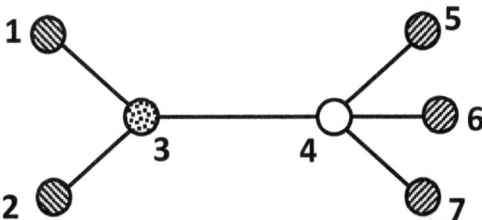

Fig. 4.1 Example of a symmetric graph (Xiao *et al.* 2008b).

4.4 Real-life networks

A *system* is a set of components functioning together as whole. Identifying a system helps us isolate a set of components which interact mainly with other members of the system, and not so significantly with what is not a part of the system.

As stated in Chapter 1, a complex system can be viewed as a *complex network*, thus enabling us to employ the full power of network theory for understanding complex systems (Newman 2003). Complex systems embody organizing principles encoded, among other things, in the topology of the underlying network.

One makes a distinction between *mathematical* networks on the one hand, and *real-life* complex networks on the other. A characteristic feature of the latter is that they usually *grow* with time, so that the number of vertices is not a constant (Albert and Barabási 2002).

Another feature of real-life networks is that they not only grow or shrink, but also *respond* to the environment, so that there is usually an element of feedback and adaptation involved.

And they have a highly heterogeneous structure.

4.5 Scale-free networks

There are many examples of so-called *scale-free* networks, in which a few of the nodes have a much larger number of connections than others. For regular networks and random networks, the number of edges per node has an approximately Gaussian distribution, with a mean value that is a measure of the *scale* of the network. By contrast, in a scale-free network there are a few strongly connected nodes and a large number of weakly connected ones. Typically, the degree distribution follows a power law:

$$P(k) \sim k^{-\gamma} \tag{4.10}$$

The exponent γ generally lies between 2 and 3. Since the distribution function $P(k)$ does not show a characteristic peak (unlike the Gaussian peak for random networks and regular networks), the network is described as scale-free (Barabási and Albert 1999; Barabási 2009).

5. Self-Organization and Symmetry

Einstein's great advance in 1905 was to put symmetry first, to regard the symmetry principle as the primary feature of nature that constrains the allowable dynamical laws. Thus the transformation properties of the electromagnetic field were not to be derived from Maxwell's equations, as Lorentz did, but rather were consequences of relativistic invariance, and indeed largely dictate the form of Maxwell's equations. This is a profound change of attitude. Lorentz must have felt that Einstein cheated. Einstein recognized the symmetry implicit in Maxwell's equations and elevated it to a symmetry of space-time itself. This was the first instance of the geometrization of symmetry. Ten years later this point of view scored a spectacular success with Einstein's construction of general relativity. The principle of equivalence, a principle of local symmetry - the invariance of the laws of nature under local changes of the space-time coordinates - dictated the dynamics of gravity, of space-time itself.

David Gross (1996)

It may appear that, because of the complexities involved, real-life networks can hardly have any symmetry. A surprise of somewhat recent origin is that *real networks are generally richly symmetric* (Holme 2006a, b; MacArthur *et al.* 2006, 2007; Xiao *et al.* 2008a, b Garlaschelli *et al.* 2010). In hindsight, this should not be surprising. It stands to reason that the evolution of real-life complex networks must be guided by some *organizing principle*, rather than being a sequence of random processes. For example, biological evolutionary and optimization forces are at play in the case of living entities.

Organizing forces and symmetry go together. Growth of a crystal (which is also a kind of network) provides an easy-to-understand example of this.

5.1 Growth of a crystal as an ordering process

Only Pierre Curie has seriously considered the causes of symmetry in Nature; some of his recorded views indicate that he saw in the symmetry of natural phenomena laws of the same generality and significance as those of thermodynamics.

A. V. Shubnikov (1944)

Consider a gas of noninteracting molecules. For it the most probable state of existence is one in which there is complete disorder in the mutual positions and velocities of the molecules. The system tends to attain this state of equilibrium, and then tends to stay there. And the entropy has its maximum value when equilibrium has been reached.

Next, let us bring in attractive interactions among the molecules. The situation changes drastically. At very high temperatures, the disordering thermal tendency dominates and we have complete disorder and maximum entropy, and equilibrium. But as temperature is reduced, condensation to a liquid state takes place at some temperature, and there is some semblance of order now.

On further cooling, the liquid may form a crystal. A crystal is a highly ordered state of matter. When a crystal grows from a fluid, matter goes from a less ordered state to a more ordered state, so there is a *local lowering of entropy*. It should be remembered that we are dealing with an *open* system here, for which the second law of thermodynamics must be stated in terms of free energy F $(= E - TS)$, rather than entropy S (Wadhawan 2010, 2017/2018). The law for open systems says that a process can occur if it lowers F. But F depends not only on the entropy S, but also on the internal energy E. So a tighter binding of molecules in a crystal compared to the fluid state can possibly result in a larger drop in E compared to the drop in the magnitude of the term TS, thus giving a net decrease in F. The state with minimum F is the state of equilibrium.

If we have to compare two competing processes for which the difference in the entropy term is not significant, we can say that that process is favoured which minimizes the internal energy E. In particular, a crystalline material *self-organizes* into that space-group symmetry which results in the least internal energy, or the maximum binding energy.

The building blocks (BBs) in a crystal are all identical. What could be the thermodynamic reason for that? It is the same as that for the fact that the molecules of a gaseous species are all identical. The crystalline material is a particular chemical species, involving a large number of molecules and the interactions among them. The asymmetric unit of the crystal can comprise of one molecule, half a molecule, or more than one molecules of this chemical species. Say it is one molecule. A particular molecule got formed and is stable because it corresponds to the largest binding energy, or the lowest internal energy. It is highly unlikely that different portions of the chemical species will settle for *different* molecular shapes or sizes, because it is most probable that only one particular molecular shape and size has the least free energy, and any other configuration therefore has a higher free energy. So the asymmetric units or molecules are identical or equal because free-energy considerations demand that.

The symmetry of a crystal is synonymous with *identical or equal placement of equal parts* (Sheftal 1966a). When these equal parts self-assemble into a crystal, it is highly unlikely that the nature of the assembly will be different in different portions of the crystal. If one portion of the crystal finds for itself (through a process of trial and error) a least-energy configuration of neighbouring asymmetric units, it is most likely that other asymmetric units will also zero-in on the same mutual configuration, with the same binding energy per asymmetric unit. Only such an arrangement can ensure that the crystal as a whole has the lowest internal energy. If different parts of the crystal were to have different arrangements of the asymmetric units, then either the asymmetric units are not identical (not possible), or the interactions among the identical asymmetric units are not the same everywhere (again not possible).

Thus, the symmetry of a crystal arises from the least-free-energy requirement imposed by the second law. The first law of thermodynamics (conservation of energy) is also involved: The tightest-binding state attained at equilibrium is the most likely state because energy is conserved. The bound molecules of the crystal cannot come apart spontaneously because the energy required to break the bonding cannot come out of nowhere.

5.2 Similar linkage patterns and symmetry

We have seen above that the maximization of binding energy (or cohesive energy), and therefore the packing density, is an organizing principle (it helps lower the free energy) when a crystal grows. And it is unlikely that this can be achieved in more than one ways for a given chemical species. This sameness of the requirement everywhere in the crystal (namely the same pattern of chemical bonding, nearest neighbours, etc.) leads to an equal placement of equal parts and the resultant symmetry. Thus, the emergence of symmetry during crystal growth is determined by the phenomenon of *growth with preferential attachment*. To borrow another phrase from the theory of complex networks, it is also a case of symmetry arising from the presence of *similar linkage patterns* (SLPs).

The challenge in complexity science is to carry over such considerations to real-life networks, with the hope that the underlying organizing principles may be decipherable from an analysis of the symmetry of the networks. One makes the fundamental assumption that the structure of a real-life network is related to its function (Holme 2006a, b). Therefore, by a study of the structural parameters characterizing a network, one should be able to learn something about the underlying forces and tendencies responsible for creating the network.

5.3 Symmetry as a secondary organizing principle

> *The ubiquity of symmetry in disparate real-world systems suggests that it may be related to generic self-organizational principles.*
>
> MacArthur and Anderson (2006)

In the example of crystal growth considered above, can we regard symmetry as an *organizing* principle? It is certainly a manifestation of the deeper organizing principle, namely the second law of thermodynamics for open systems. Crystalline order emerges spontaneously from the highly disordered fluid state because its emergence results in the least possible free energy locally: The binding-energy contribution to the lowering of the free energy overpowers the smaller local decrease in entropy. A symmetrical arrangement of atoms in an open system has the lowest free energy (Fig. 5.1), which means that, given the right conditions, symmetry can indeed be a manifestation of the basic underlying self-organization principle, namely the second law.

Fig. 5.1 has a rectangular unit cell, but there is really no directional symmetry present. Its plane group is p1. By contrast, the crystal structure in Fig. 5.2 does have a 2-fold axis of symmetry, and its plane group is p2. Symmetry is these two examples is incidental to the consequences of the basic underlying organizational principle. All phenomena must obey the second law of thermodynamics. For thermodynamically open systems this means that the free energy F ($= E - TS$) can never increase; it must either decrease or remain constant ($\Delta F \leq 0$). The translational and directional symmetry of crystals is certainly a consequence of the second law, as is every other process in Nature.

The second law for open systems is the overarching organizing principle for all phenomena, including the emergence of all kinds of symmetry.

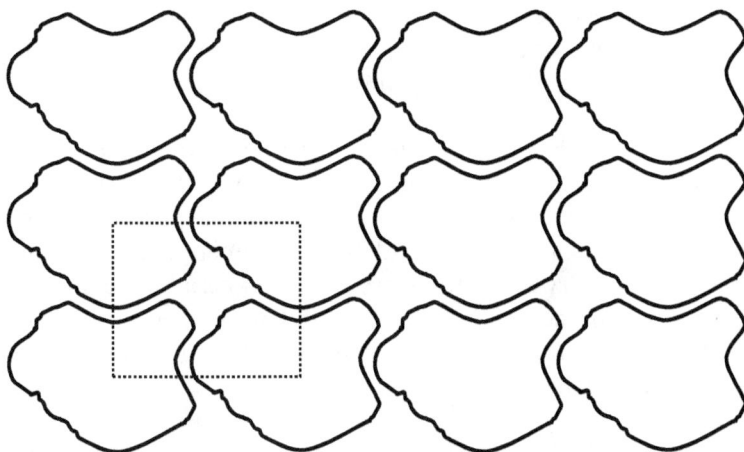

Fig. 5.1 Shown here is an equal placement of equal parts, namely the molecules of a highly asymmetric organic substance. Translational symmetry ensues from the tendency of the molecules to minimize the overall binding energy by the fitting of the projections of every molecule into the hollows of neighbouring molecules. The dotted-line rectangle is the unit cell for this schematic 2-dimensional crystal; but the structure has no symmetry other than translational symmetry.

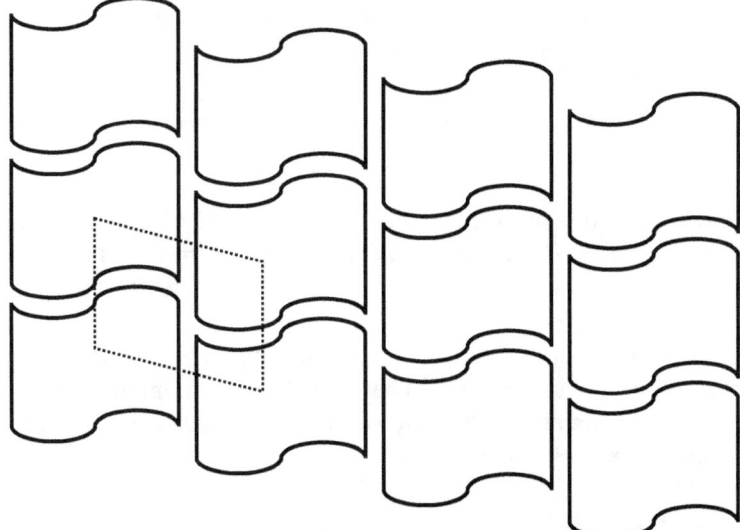

Fig. 5.2 A schematic 2-dimensional crystal which not only has translational symmetry, but also rotational symmetry, namely a 2-fold symmetry axis. The unit cell in this case is a parallelogram.

The second law is the *primary* organizing principle for everything that occurs in nature. And symmetry is a major consequence of the second law.

So much so that there is a case for regarding symmetry as a *secondary* self-organizing principle for open systems: It ensures a minimization of the free energy. Here are some instances of this:

1. When a crystal is growing, the translational-symmetric bonding of molecules helps maximize the magnitude of the overall binding energy, and therefore minimizes the free energy. Even the minimal placement-symmetry present in every crystal, namely translational symmetry, helps minimize the free energy. Symmetry is the necessary condition here for the minimization of free energy. Any deviation from the spontaneously acquired placement symmetry would result in incomplete short and long chemical bonds, or dangling bonds (Sheftal 1966b).

2. Many real-world complex networks are richly symmetric. In some of these cases, the 'branching' occurring as the network grows (by the preferential attachment of new nodes) runs concurrently with the evolution of symmetry (MacArthur and Anderson 2006b). Rather like what happens when a crystal grows.

3. Preferential attachment in certain complex networks also enhances tolerance to attack (Albert, Jeong and Barabási 2000), and is thus an organizing principle in the evolution of certain living organisms (just as preferential attachment of molecules during crystal growth is an organizing principle, running concurrently with the emergence of crystallographic symmetry).

4. The regular and repetitive chemical bonding in a crystal is an example of repetition of *similar linkage patterns*. The phrase 'similar linkage patterns' has been used recently by Xiao *et al.* (2008a, b) while describing the emergence of symmetry in real-life complex networks. Why do similar linkage patterns occur in such networks? They do so for the same reason for which they occur in crystals: They result in maximum cohesiveness (chemical cohesiveness in the case of crystals; social cohesiveness in the case of social networks; and so on).

5.4 Symmetry and biology

> *The most general law in nature is equity - the principle of balance and symmetry which guides the growth of forms along the lines of the greatest structural efficiency.*
>
> Herbert Read

Most biological systems possess a substantial amount of approximate symmetry which, at first thought, appears surprising. They are highly complex conglomerations. Why should they possess any symmetry at all? Dawkins (1996) devoted considerable attention to this question, and even today investigations of this aspect of evolutionary biology are common (Wolynes 1996; Andre *et al.* 2008).

Since most of the biological forms, particularly the land-based ones, have substantial symmetry, there must be an evolutionary advantage to it. Mutations can lead to changes of body shape, and this usually happens through an adjustment of the processes of *embryonic growth*. Which mutations become favourable for the species depends on the nature of the embryology involved. For example, an embryology may be mirror-symmetric, or it may not be mirror-symmetric, and the question is: Which one is more evolvable in a given set of conditions?

Dawkins (1996) gave the analogy of the kaleidoscope to answer this question. A kaleidoscope has a set of mirrors which generate a systematic (symmetrical) pattern even from a *random* heap of coloured chips. Although mutations are like random taps on the positions of the chips (in the genotype), what we see are the changes as influenced by the mirrors of the kaleidoscope (if any such mirrors are present in the phenotype). Symmetrical structure can exist in the final product (phenotype) even though none exists in the slight and random mutations, or in the changes in the positions of the chips.

[It may be mentioned parenthetically that any *latent* symmetry in the chip positions can result in *additional* symmetry, and perhaps additional evolutionary advantages.]

It has been argued that a symmetrical (spatially repeating) pattern of embryology has evolutionary advantages. How?

Most mutations are no good, and just do not propagate in the species. Natural selection takes care of that. But if there is a beneficial mutation, its effect on the embryology can get multiplied if the embryology involved has symmetry. For example, if there is left-right (mirror) symmetry, a single good mutation will have *twice* the good effect because both the left and the right part of the body will benefit. In the absence of such a symmetry, natural causes will have to improve the left and the right parts independently, which may never happen, or may happen after a huge lapse of time. Thus, the emergence of symmetry can speed up evolution. Mutations are random, but natural selection is non-random. An extra element of nonrandomness is brought in by the presence of symmetry. And there is an *amplifier effect* too: The effect of any good mutation can get repeated or multiplied at many (symmetrically equivalent) places in the body (Monod, Wyman and Changeux 1965).

Natural selection operates at the genotypic level, and the effects are seen at the phenotypic level. Even a cursory look reveals the *need* for certain symmetries, at least in terrestrial animals. Because of gravity, top-bottom symmetry cannot be there. Similarly, front and back cannot be symmetrically related. For example, food enters from the front and excretion occurs from the back. Similarly, the need to run away effectively from danger must make a distinction between forward and backward. But left-right biological symmetry makes a lot of sense. Imagine a creature with no left-right symmetry. It would end up moving in circles, instead of moving straight ahead and away from a predator. In any case, as Dawkins (1996) pointed out, there is no reason why left-right symmetry should *not* be there. Instead of saying 'why?', we may as well say 'why not?'.

Thus if some mutation is good for the left, it is good for the right too. In the beginning the left and the right were different, and only a long natural-selection process could make them identical. But once that happened, evolutionary advantage ensured that it (the left-right symmetry) propagated preferentially in the gene pool. This is how symmetry is built into the genetic make-up of the future generations. And this is how *additional* symmetries could have developed. Mostly it was a one-way traffic. Once a symmetry emerged, it was unlikely that it would be lost on further evolution.

For creatures living in the sea the effective downward pull from gravity is greatly reduced. And swimming, rather than walking, allows for an easy rolling-over of the body. Thus, although top-bottom asymmetry is still present, a forward-backward mirror symmetry becomes less unviable than on land. And, if high symmetry is indeed a good thing for evolution and evolvability, why stop at just two planes of mirror symmetry? This seems to be the logic or rationalization for the remarkably high symmetry evolved by stalked jellyfishes (Fig.5.3).

Dawkins (1996) also pointed out the widespread existence of translational symmetry in Nature. This so-called '*segmentation*' can be seen in, for example, centipedes. There is a linear repetition of a certain module, except near the front and the rear.

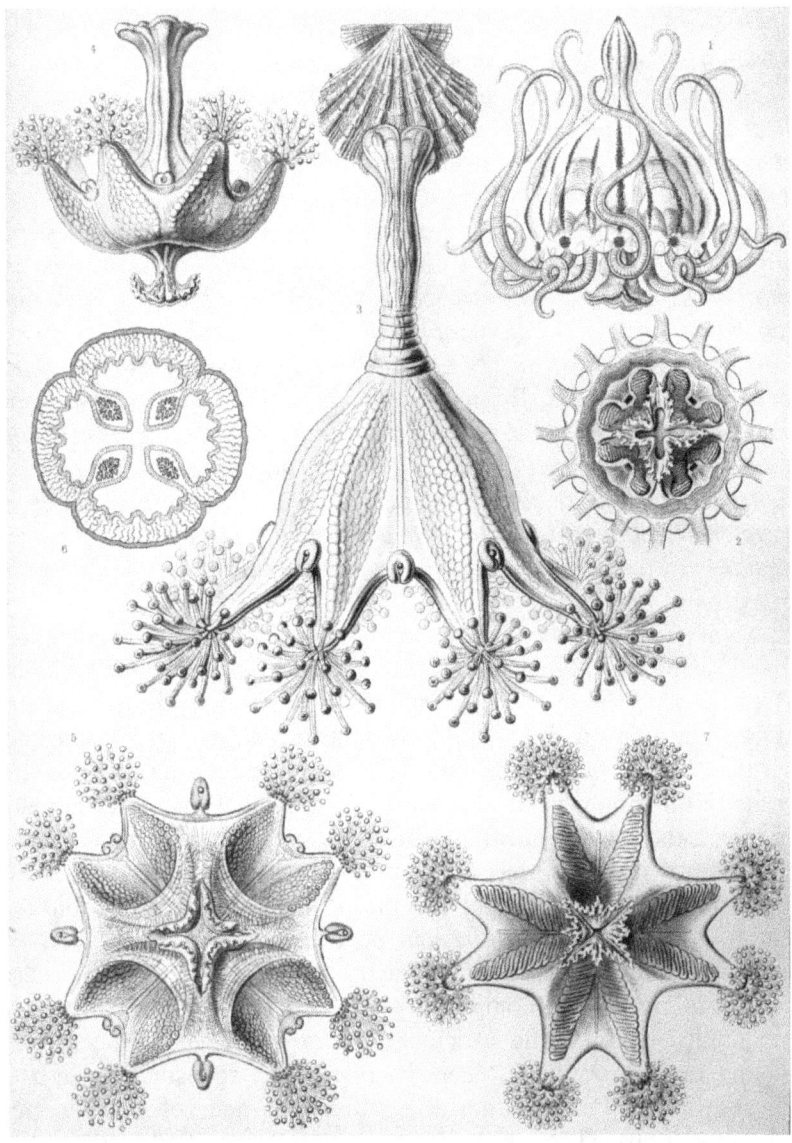

Fig. 5.3. Stalked jellyfishes. It is tempting to speculate that latent symmetry must be playing a role in the emergence of such high symmetries in biological systems.
[http://commons.wikimedia.org/wiki/File:Haeckel_Stauromedusae.jpg]

I quote from Dawkins (1996) to convey the essence of the evolutionary underpinning of symmetry in biological forms:

'The central message of this chapter is that kaleidoscopic embryologies,

whether working through segments and clusters of segments arranged in a line from front to rear as in an insect, or through 'mirrors' of symmetry as in a jellyfish, are paradoxically both restrictions and enhancements. They restrict evolution in that they limit the range of variation available for selection to work upon. They enhance evolution in that - to put it in language that personifies selection forgivably - they save natural selection from wasting its time exploring vast regions of search space which are never going to be any good anyway. The world is populated by major groups of animals - arthropods, molluscs, echinoderms, vertebrates - each one of which has a form of kaleidoscopically restricted embryology which has proved evolutionarily fruitful. Kaleidoscopic embryologies have what it takes to inherit the earth. Whenever a major shift in kaleidoscopic mode or 'mirror' has spawned a successful evolutionary radiation, that new mirror or mode will be inherited by all the lineages in that radiation. This is not ordinary Darwinian selection but it is a kind of high-level analogy of Darwinian selection. It is not too fanciful to suggest as its consequence that there has been an evolution of improved evolvability'.

Even a simple thermodynamic approach provides easy rationalization of symmetry in biology. Phenomena occur because their occurrence minimizes the free energy F ($F = E - TS$). Crystals grow because in them the strong binding of the atoms in the crystal lowers the internal energy E by an amount larger than the decrease in the magnitude of the entropy term TS due to the ordering of the arrangement of atoms in the crystal. Thus, order can indeed emerge in a thermodynamically open system under the right conditions. A similar line of reasoning has been articulated by Andre *et al.* (2008) for rationalizing the emergence of symmetry in homooligomeric biological assemblies. There have been various speculations about the evolutionary advantages of the high degree of symmetry possessed by naturally occurring homooligomeric protein complexes (Goodsell and Olson 2000): Greater 'designability' and folding efficiency of the proteins; increased coding economy; greater stability; reduced aggregation; robustness to errors in synthesis; amenability to allosteric regulation; and, of course, greater adaptability discussed above. Andre *et al.* (2008) argue that that mutation is more likely to survive in the gene pool which results in a *stronger binding* among the subunits of the protein. And this requirement also happens to serve the purpose of making the internal-energy term E dominate over the entropy term TS in the expression for the free energy. This is how the germs of symmetry are sown, because a domination over the entropy term means emergence of order and symmetry, like in crystal growth.

A large number of proteins can be crystallized, although the crystal size is usually small. An interesting observation in many cases is that the asymmetric unit of the crystal contains more than one molecules of the protein. These subunits within the asymmetric unit can have a symmetry relationship which is not reflected in the space group of the crystal, and we then speak of *non-space-group symmetry* (Fichtner 1986). Why does it occur at all? For free-energy-minimization reasons, of course. In such cases the shape of the protein is such that a single-molecule asymmetric unit does not result in a high lowering of the local free energy. Instead, a superunit, namely an asymmetric unit comprising of two or more molecules, acquires a shape and force field which can result in a crystal with a stronger binding energy, and often a higher symmetry.

6. The Different Types of Exact and Approximate Symmetry

It is instructive to try to enumerate or classify the different types of symmetry. Let us begin by considering the symmetry of crystals.

6.1 Crystallographic symmetry

The essence of the symmetry of a crystal lies in the manifestation of *equivalence* among subparts of the crystal (Sheftal 1966a, b, 1976; Vainshtein and Chernov 1988; Lederman and Hill 2004/2008). Because of this equivalence, when certain transformations (called 'symmetry transformations') are applied, the crystal transforms back into itself. A crystal is thus a composite system made up from its equal or equivalent subparts.

It is not sufficient that a symmetric object be a collection of equal parts. An equal or identical *placement* of the equal parts is also necessary. A jumble of unit cells will not constitute a crystal. The unit cells must lie on a *lattice* of repetitions. A jumble of unit cells or other equal objects may have symmetry only in a statistical sense. Thus the symmetry of a crystal arises from *an equal or identical placement of equal parts* (Sheftal 1976). And this *placement symmetry* (PS) is an important ingredient of the symmetry exhibited by many other objects, and not just crystals.

The full (non-magnetic or 'chemical') symmetry of a crystal is described by one of the 230 crystallographic space groups. These space groups also describe exhaustively all possible *site symmetries* in crystals. Site symmetry is essentially a directional symmetry, so it must be one of the 32 crystallographic point-group symmetries (Burns and Glazer 1990; Wadhawan 2000).

The notion of the *asymmetric unit* is familiar in crystallography. It is that smallest portion of the unit cell such that the full crystal can be generated from it by applying all the symmetry operations of the space group of the crystal. Any such (nontrivial) space-group operation does not map the asymmetric unit back onto itself.

6.2 Space symmetry and time symmetry

> *With the development of quantum mechanics in the 1920s symmetry principles came to play an even more fundamental role. In the latter half of the 20th century symmetry has been the most dominant concept in the exploration and formulation of the fundamental laws of physics. Today it serves as a guiding principle in the search for further unification and progress.*
>
> David Gross (1996)

At a basic level one can distinguish between space symmetry and time symmetry, although the combined space-time symmetry can also become important in certain situations (Ananthaswamy 2010).

In the context of symmetry-breaking in 'dissipative structures', Prigogine (1977) argued that the appearance of any periodic motion is a time-symmetry-breaking process, just as the emergence of a ferromagnetic phase at a phase transition in a crystal is an instance of space-symmetry breaking.

In studies of the emergence of space-symmetry breaking, the focus is on the changes occurring in the space in which the system is embedded; consequently, the changes are generally taken to be of an *'exogenous'* nature. By contrast, emergence of time-symmetry breaking is relevant in the context of evolution of dynamical systems. One considers changes that are *endogenous* to the system, with space providing a constant, homogeneous backdrop. Time-symmetry breaking has received much attention in the context of evolution of highly nonlinear complex systems. By contrast, it is only recently that breaking of *space* symmetry in complex networks has received substantial attention (Ruzzenenti *et al.* 2010).

6.3 Permutational and more general symmetries of graphs

A graph may have discrete symmetry, rather like the discrete translational symmetry of a crystal lattice. It is interesting that the discrete translational symmetry of a crystal lattice also makes its rotational symmetry discrete: Only 2, 3, 4, and 6-fold rotational symmetries are possible, either alone or in conjunction with one another, and/or with inversion symmetry. It is conceivable that, similarly, the discrete symmetry of graphs may impose restrictions on other possible symmetries of the graphs.

Unlike crystal lattices, graphs are *topological* entities, rather than *geometrical* ones. We may *represent* a graph by locating its vertices in some metric space, but its properties are independent of the locations chosen for the vertices. Changing the positions of the vertices only changes how a graph *looks*, but has no effect on its topology or the connections among the vertices.

We can identify each vertex by giving it a label; e.g. 1, 2, 3, ... A *topological transformation* is one which maps each vertex to a vertex determined by its identity or label (and not by its coordinates in the representation space). It is thus merely a *permutation* of the vertices.

An *automorphism* of a graph is a permutation of its vertices that leads to the same topology as before. It is thus a symmetry operation.

The vertices of a graph may have other properties than what is reflected in just its topology. It happens often in real-world situations that one knows only the topology of the complex network, and not the inherent properties of every vertex. In such cases, automorphism operations are defined by ignoring the intrinsic properties of the vertices.

Vertex permutations are not the only way under which a graph may possess symmetry. Self-similar or fractal objects may exhibit a more general type of symmetry, namely *scale invariance*: In this case the symmetry transformation is a change of scale. There are other possibilities also, which we shall consider below.

6.4 Approximate symmetry of graphs

Unlike in abstract graphs, the symmetry of graphs underlying real-life networks is only an approximate symmetry. A geometrical analogue would be the symmetry of a mathematically perfect circle *vs.* the symmetry of a real circle. It is impossible to draw a perfect circle in real life. And yet it would be inadvisable to ignore the symmetry of a real circle. One therefore introduces the notion of approximate symmetry. A perfect circle is invariant under any rotation about its centre and in its plane. But for a real circle, any such rotation will take the points on the circle to points which lie *near* the existing points on the circle. It follows that there are infinitely many imperfect circles of a given average radius, and we end up with a *family* of circles, rather than a single unique circle.

As discussed by Garlaschelli *et al.* (2010) and Ruzzenenti *et al.* (2010), any real network, as also any object or entity having imperfections and errors, requires a stochastic notion of symmetry. They have therefore proposed a definition of *stochastic symmetry* of real networks based on *ensembles* of the underlying graphs. They have demonstrated that it is stochastic symmetry which highlights the most informative topological properties of real networks. And this is true even for noisy situations, inaccessible to exact analyses.

To describe the approximate symmetry, one needs the entire ensemble of variants. For an imperfect circle, for example, a rotation maps it to a different imperfect circle. One can associate a probability with each approximate object in the ensemble, resulting in a *statistical ensemble* of objects. The probability associated with each member is proportional to how close it is to the perfect object. Further, objects different from the perfect one by the same amount are equiprobable to occur. Approximate symmetry is said to exist in a given real object if it is a '*typical*' object; i.e., if its probability of occurrence in the ensemble is high.

Thus, for real networks one can define a statistical ensemble of the underlying graphs. Each graph G has a probability of occurrence $P(G)$.

A graph is *exactly symmetric* under a transformation if it is mapped onto itself by the transformation. And it is *stochastically symmetric* if it is a typical member of a graph ensemble which is stochastically symmetric under that transformation: A stochastically symmetric graph ensemble is one for which, under a given transformation, a member graph G_1 is mapped onto another *equiprobable* member graph G_2; i.e., $P(G_1) = P(G_2)$.

6.5 Symmetry in real-life networks

Near-absence of translational symmetry

It can happen sometimes that the graph underlying a network is *embedded* in some metric space, instead of being just a topological entity. The automorphisms of such a graph are then coordinate transformations of the vertices. A crystal lattice is clearly an example of this. In it, connections of vertices to their nearest neighbours construct the entire lattice. And it has translational symmetry (at least). *Crystal lattices are a particular case of regular graphs*: every vertex has the same number of edges.

But practically all real-life networks and graphs violate translational symmetry. An example is graphs exhibiting the so-called *small-world effect* (Caldarelli 2007). In such graphs the average number of links needed to traverse along the shortest path connecting any two vertices is found to be rather small. This has a bearing on, for example, the way epidemics spread in a population. The spreading rate is far higher than what one would expect if translational symmetry were nor violated so strongly.

Scale invariance

Real networks tend to be topologically heterogeneous. The degree distribution has a power-law form:

$$P(k) \propto k^{-\gamma},\tag{6.1}$$

with $2 < \gamma < 3$. This means that there are many vertices that have very few links, and only a few vertices (*hubs*) that have a large number of links. Such networks are described as *scale-free* networks because, unlike in a Gaussian distribution (in which the mean value defines the typical scale of the distribution), here there is no characteristic path-length scale or the most commonly occurring value for the degrees of the vertices.

Scale-free networks possess a type of symmetry called *scale-invariance symmetry*: The topology looks the same at various length scales, here 'length' meaning the shortest path length between two vertices. Suppose the scale of observation is changed by switching from degree k to degree αk, where α is a positive number. Then the number of vertices with a given degree changes only by a magnification factor, from $P(k)$ to $P(\alpha k) = \alpha^{-\gamma} P(k)$. In other words, the same power-law distribution holds.

The symmetry group associated with scale-invariance symmetry is called the *renormalization group* (Stanley 1999).

Scale-invariance symmetry is widespread in real networks, and points to the presence of some common underlying principle governing the emergence and growth of these networks. A pioneering piece of modelling of this phenomenon was carried out by Barabási and Albert (1999). They introduced two basic postulates in their model (now called *the BA model*):

(i) Real networks *grow* with time, meaning that new vertices keep adding to the existing network. This is analogous to the growth of a crystal.

(ii) If some vertices are highly connected compared to others, there must be a reason for this, and the same reason may continue to hold, meaning that there is a higher probability that a new vertex will link with a highly connected vertex rather than with a less connected one (*'rich getting richer'*). This is *not* analogous to what happens in crystal growth, wherein the saturation of chemical bonding, as also factors like steric hindrance, discourage the 'rich getting richer' phenomenon, and instead result in translation symmetry. [However, as highlighted by the work of Xiao *et al.* (2008b), there is a widespread occurrence of 'similar linkage patterns' (SLPs) in growing real-life networks. In my opinion, this does correspond to the obedience of rules of chemical bonding during the growth of a crystal. I shall elaborate on this in Sections 13.4 and 13.5.]

The 'rich getting richer' syndrome in the growth of a real network is also an example of *preferential attachment*. The Barabási-Albert model says that, if k is the degree of an existing vertex, the probability that a new vertex will be attached to it is $P(k) \sim k^{-\gamma}$. This power-law distribution, in contrast to a Gaussian distribution, implies the absence of a characteristic length scale. This is the essence of a scale-free network.

Graph automorphism and structural invariance

Consider a graph in which each vertex has a distinct label; we call it a *labelled graph*. The graph may be invariant under certain permutations of vertices. Such operations are automorphisms. The group of all such automorphisms is called the *automorphism group* of the graph.

Automorphism groups of real networks are of great topical interest. One reason for the current interest is the important notion of '*structural equivalence*' in graph theory. Suppose two vertices i and j in a labelled graph have exactly the same set of neighbours. Then a permutation which interchanges only these two vertices results in exactly the same graph topologically. In social science the term *structurally equivalent* is used for the vertices i and j (Wasserman and Faust 1994). Fig. 6.1 shows an example of a graph having structurally equivalent vertices.

Structural equivalence is important from the point of view of *robustness* of real networks. Suppose the structurally equivalent vertices i and j mentioned above are major hubs in the network. Then even the total decimation of one such hub need not have serious qualitative consequences because the other equivalent hub is intact to carry on the activities of the network. But if there is only one such hub and it gets eliminated, the consequences for the topology can be disastrous; such a network is not robust.

As a real network grows and evolves, the emergence of more and more structurally equivalent vertices is likely to occur, rather like the emergence of the same atomic structure and symmetry in different parts of a growing crystal. Structural equivalence and redundancy is thus a type of symmetry in real networks, and throws light on the processes involved in the evolution of a network.

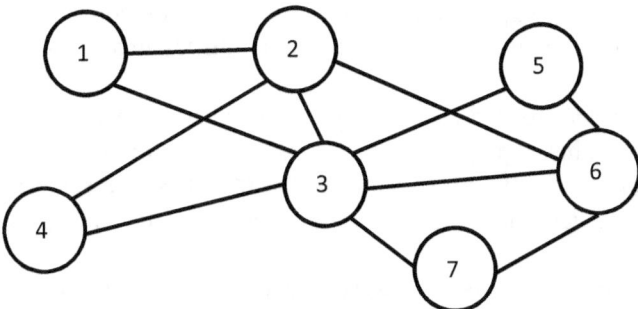

Fig. 6.1 A graph with some structurally equivalent vertices. Vertices 1 and 4 are structurally equivalent. Interchanging them is a symmetry operation (an automorphism). Since both are connected to 2 and 3 only, interchanging them does not affect the topology of the graph. Similarly, vertices 5 and 7 are structurally equivalent; both are connected to 3 and 6 only. [After Garlaschelli *et al.* (2010).]

Recent work on automorphism in real networks has revealed that, unlike random networks, real networks contain a substantial amount of *structural redundancy* (Alon 2003; MacArthur *et al.* 2008, 2009; Xiao *et al.* 2008c; Wang *et al.* 2009). This has been correlated to the processes involved in the growth and evolution of real networks.

Network motifs (NMs) are striking examples of structural invariance in a variety of biological circuits (Alon 2003, 2007; Wadhawan 2007). Although mutations are random, natural selection is not. NMs are circuit elements which have been discovered again and again by biological evolution. Each such motif in the molecular network performs a specific information-processing job. They are like operational amplifiers and memory registers in human-made electronic circuits.

Formally, NMs are defined as patterns of interconnections occurring in biochemical complex networks in numbers significantly larger than in the corresponding randomized networks (Milo *et al.* 2002; Shen-Orr *et al.* 2002).

Statistical equivalence

Statistical equivalence of vertices is to structural equivalence of vertices what approximate symmetry is to exact symmetry. Since real networks have only approximate or stochastic symmetry, an analysis of statistical equivalence can be a richly rewarding experience in discerning the symmetry of large real networks. Here are some examples of the type of questions one attempts to answer in an investigation of the average topological equivalence, or invariance, or symmetry in a real network:

• Do all vertices have the same degree, irrespective of the identity of their neighbours? This type of symmetry is usually not present (not even approximately) in real networks because of their frequent scale-free nature.
• Do the vertices have the same number of second neighbours, on an average?
• Does the network have vertices whose neighbours have the same average degree, irrespective of the number of neighbours of each vertex?

A number of *statistically* equivalent vertices and symmetries may exist in a real network, even though there may be no set of strictly equivalent vertices. The structure of the statistical equivalence classes determines the symmetry of the network under investigation.

Invariance under permutation of 'external' properties

One would like to trace back the observed complexity of a network (including any symmetry exhibited by it) to some fundamental nontopological properties involving the '*hidden*' *variables* attached to the vertices (Bianconi *et al.* 2009; Garlaschelli *et al.* 2010). Fig. 6.2 illustrates what this can mean. It illustrates a model for scale-free networks suggested by Caldarelli *et al.* (2002).

A fitness value x_i is assigned to every vertex in accordance with some probability distribution function $\rho(x)$. Then links are drawn among all pairs of vertices with a probability depending on the fitnesses of the two vertices in the pair. This gives rise to a '*good get richer*' mechanism in which vertices with high fitness values are more likely to become hubs, thus providing an explanation for the scale-free nature of such networks. And permutation symmetry arises when for two vertices i and j, $x_i = x_j$.

Another important example of 'hidden variable' or 'external variable' or 'fitness' is what Xiao *et al.* (2008b) called the '*initial degree*' of any vertex of a growing network. I shall discuss it later in Chapter 13.

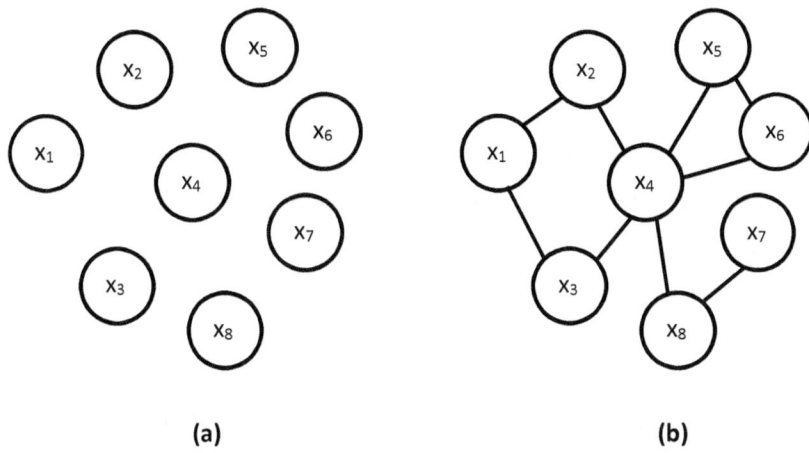

(a)   (b)

Fig. 6.2 Invariance under permutation of hidden variables attached to the vertices of a net. Such a variable can be, for example, the intrinsic fitness x_i associated with each vertex i. (a) In the fitness model proposed by Caldarelli *et al.* (2002), one begins with an unconnected set of vertices, each assigned a fitness value drawn from some specified distribution function $\rho(x)$. (b) Links between any two vertices x_i and x_j are drawn with probability defined by a connection function $p(x_i, x_j)$. Permutation symmetry arises because any two vertices with the same fitness value are statistically equivalent. Because of this statistical equivalence, all the topological properties of the two vertices have the same expected values. [After Garlaschelli *et al.* 2010.]

Ensemble equiprobability

Important symmetries associated with statistical ensembles of graphs may exist. Therefore, if an ensemble is a good model of a real network, the symmetries observed in the ensemble can prove valuable in gaining insights into the real network. A statistical ensemble of graphs is a collection of M graphs $\{G_1, G_2, ..., G_M\}$, each with an associated occurrence probability $P(G)$ satisfying the condition (Park and Newman 2004; Garlaschelli *et al.* 2010)

$$\sum_G P(G) = \sum_{m=1}^{M} P(G_m) = 1 \tag{6.2}$$

One would like to ensure that the model ensemble of graphs is maximally random (under the given set of constraints). Therefore, the distribution function for the probabilities $P(G)$ should be such that it maximizes the Shannon-Gibbs entropy

$$S \equiv -\sum_G P(G) \ln P(G) \tag{6.3}$$

The entropy maximization is done under the set of enforced constraints, which are a collection $\{c_1,....,c_K\}$ of K topological properties, forming a K-dimensional vector $\vec{c}$:

$$\vec{c} = \{c_1,.....,c_K\} \tag{6.4}$$

Each property $c_a (a = 1,..., K)$ materializes as $c_a(G)$ when measured on the particular graph G.

The entropy maximization can be done in various ways, depending on whether one works with a microcanonical ensemble, a canonical ensemble, or a macrocanonical ensemble (Garlaschelli *et al.* 2010). In each case one works out the graph Hamiltonian $H(G)$ and the partition function Z. For example, for a canonical ensemble, one requires that the constraints $\vec{c}$ be met *on average*. This gives

$$P(G) = \frac{e^{-H(G)}}{Z} \tag{6.5}$$

Here $H(G)$, the graph Hamiltonian, is a linear combination of the constraints:

$$H(G) \equiv \sum_{\alpha=1}^{K} \theta_\alpha c_\alpha(G) \tag{6.6}$$

The partition function Z is defined by

$$Z = \sum_G e^{-H(G)} \tag{6.7}$$

The Hamiltonian $H(G)$ is a measure of the *energy* or *cost* associated with a given network configuration, and determines uniquely the probability $P(G)$.

Consider two graphs G_1 and G_2. They have the same ensemble probability $P(G_1) = P(G_2)$ if

$$H(G_1) = H(G_2) \tag{6.8}$$

This means that the symmetries of $H(G)$ are transformations that map any graph G_1 in the ensemble to an equiprobable graph G_2. G_1 and G_2 have different topologies but the same values of the enforced constraints. This is stochastic symmetry. Such symmetry operations map the graph onto a different one in the same statistical ensemble, rather than back onto the same graph.

Real networks have imperfections and errors, unlike mathematical networks generated by deterministic algorithms. Therefore, only a stochastic notion of symmetry can be applicable to real networks. Such recognition and analysis of approximate symmetry can be very informative, as it can discern order even in intrinsically noisy situations (Ruzzenenti *et al.* 2010).

6.6 Structural *vs.* statistical equivalence and latent symmetry

Fig. 6.2 above provides a good example of the distinction between structural equivalence and statistical equivalence. Vertices 2 and 3 are structurally equivalent: both are connected to 1 and 4 only. But they are not statistically equivalent because $x_2 \neq x_3$.

When $x_2 \neq x_3$, one would expect that, in general, the number of linkages (i.e. the degrees) of the vertices 2 and 3 would be different. When $x_2 \approx x_3$, the possibility of the degrees of 2 and 3 being the same becomes more likely, and this is what is depicted in Fig. 6.2.

Permutation symmetry associated with structural equivalence is akin to placement symmetry in a crystal. *Is it that the permutation symmetry associated with statistical equivalence ($x_2 = x_3$ in the example considered here) is akin to latent symmetry?* Perhaps yes. It is conceivable that the hidden or intrinsic variables associated with a vertex, if they are also the same for another vertex in the network, result in a manifest symmetry of the network which is over and above the placement symmetry just mentioned.

Each of the two equivalent vertices by itself may not have any manifest symmetry. But when two of them interact in a network, the symmetry inherent (latent) in the vertices may manifest itself if the interactions and topology happen to be of a nature conducive to the manifestation of the latent symmetry.

Fig. 6.2 illustrates the model put forward by Caldarelli *et al.* (2002) for understanding the occurrence of a class of scale-free complex networks. The essential idea is that if some vertex has a high level of fitness, it would form more linkages with other vertices compared to lower-fitness vertices. The idea clearly has the germ of hub formation, and the attendant heterogeneity of linkages characteristic of several scale-free networks. Let us imagine such a network having two hubs, 1 and 2, of comparable fitness and degree. Fitness is something that varies as time passes. Let $x_1 > x_2$ in the beginning. This means that 1 is a stronger hub than 2 to start with, so there is a stronger chance that any new or existing vertex will link to 1 rather than to 2. Suppose x_2 increases with time, and $x_1 = x_2$ at some stage in the history of the network. Now there is no clear statement possible about the linkage preference of a new or existing vertex in the network. What we have here is the equivalent of an 'instability' in the vicinity of a phase transition in a crystal (see, e.g., Wadhawan 2000).

Let us concretize our discussion to the arena of evolution of a species, as also the emergence of a new species (speciation). Even when $x_1 = x_2$, the fact remains that 1 and 2 are different individuals or vertices, each with its own set of surroundings, neighbours, and interactions. So their evolutions will still be different. Since both are equally fit hubs, even a minor barrier to intermingling of the individuals linked to them separately can result in possible speciation. So the instability or the phase-transition point at $x_1 = x_2$ can be a possible *bifurcation point* (in phase space) in the phylogenetic tree. This bifurcation point is the *common ancestor* for the two branching or diverging species (Dawkins 1996).

7. Symmetry of Composite Systems

Modern scientists often share with the Pythagoreans of Antiquity the belief in a cosmos ordered and in balance by the highest and most perfectly mathematical laws: in the beginning, there was symmetry and simplicity.

Klaus Mainzer, *Symmetry and Complexity*

There are hardly any systems which operate in isolation; perhaps none. There is always the influence of other systems. What happens to the symmetry when two or more systems are superimposed, or when there is some interaction among them? This question is best answered in terms of the Curie principle of superposition of symmetries, or rather superposition of *dissymmetries* (Curie 1884, 1894). We had a glimpse of this principle in Chapter 2.

7.1 The Curie principle

Suppose there are two superimposed systems with symmetries described by groups G_1 and G_2. Common sense says that if some symmetry element is present in one of them but not in the other, then it would not survive in the symmetry group of the combined or composite system. Such considerations led Curie (1894) to enunciate his celebrated symmetry principle, which I outlined in Section 2.6. I repeat here the three-part statement of the principle:

1. When several phenomena of different origin are superimposed in one and the same system, their *dissymmetries* are summed. There only remain the symmetry elements common to each phenomenon taken separately.

2. When certain causes lead to certain effects, the symmetry elements of the causes should be observed in these effects.

3. The statement contrary to these two conclusions is wrong, at least in practice; that is, the effects may be more symmetrical than the causes.

The word 'dissymmetries' used above needs to be explained. Suppose G is the symmetry group of a geometrical or physical object. Dissymmetry (D) means 'absence of symmetry'. For specifying it, one must first introduce a

fundamental group or *embracing group* $\tilde{G}$ for the problem under consideration (Shubnikov and Koptsik 1974; Koptsik 1983). For example, if one is dealing with the classical tensor properties of nonmagnetic perfect crystals, then it is the group of all rotations: $\tilde{G} = O(3)$. The dissymmetry is defined as the complement D of the group G: $D = \tilde{G}/G$. That is, the dissymmetry is that part of $\tilde{G}$ that is not in G. Fig. 7.1 illustrates this.

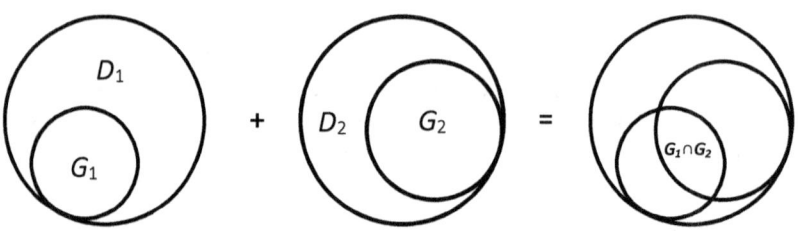

Fig. 7.1 Venn-Euler diagrams illustrating the Curie principle. The big circle represents the domain of the embracing group $\tilde{G}$. The dissymmetries D_1, D_2 etc. add up when the objects are superimposed, and the symmetry that survives for the composite object is the intersection symmetry $G_1 \cap G_2 \cap \dots$.

When the dissymmetries add up, the symmetry that survives in the composite system is what is common to all the component parts. As can be seen from Fig. 7.1, since

$$D_1 + D_2 \dots = \tilde{G}/G_1 \cup \tilde{G}/G_2 \cup \dots, \tag{7.1}$$

the net symmetry, G_d, of the composite object or phenomenon is the highest common subgroup of G_1, G_2, .. It is an intersection group obtained by taking due account of the mutual disposition and orientation of the symmetry elements of the groups involved:

$$G_d = G_1 \cap G_2 \dots = \cap_i G_i \tag{7.2}$$

Naturally, G_d cannot be higher than any of G_1, G_2, ... etc.:

$$G_d \subseteq G_i, \qquad i = 1, 2, \dots \tag{7.3}$$

Eq. 7.3 embodies the *Neumann principle* of crystal physics (cf. Nye 1976; Wadhawan 2000). According to it, *the point-group symmetry (G_i) possessed by any macroscopic physical property of a crystal cannot be lower than the point-group symmetry (G_d) of the crystal.*

The subscript d in Eqs. 7.2 and 7.3 stands for *dissymmetrization*, or symmetry-lowering.

Although Eq. 7.3 was deduced from the Curie principle, one can as well start from Eq. 7.3 and deduce the Curie principle.

Moreover, out of the set of three connected statements by Curie, which I gave at the start of this section to enunciate the Curie principle, the last two enable us to state the principle in a form which emphasizes its cause-effect aspect (discussed in Chapter 2):

The symmetry group of the cause is a subgroup of the symmetry group of the effect.

Mathematically,

$$G_{cause} \subseteq G_{effect} \tag{7.4}$$

7.2 The Curie-Shubnikov principle

The Curie principle described above does not cover all situations, and needs to be generalized. This generalization was given in the great book by Shubnikov and Koptsik (SK) (1974), and I outline it in this section.

SK gave an interesting account of the history of the Curie principle. For another historical account, see Brandmuller (1986). Chronologically, it was F. E. Neumann (1798-1895) who first realized that there is a relationship between the structure of a crystal and its physical properties. I have stated above the well-known Neumann principle of crystal physics (see Nye 1976). In 1884 Minnigerode published an 'empirical principle' which is better known today as the Neumann principle. He stated it as follows:

The group of the structure of a crystal is contained in the group of each of its physical properties.

On 13 November 1884 P. Curie (1859-1906) gave his now-famous lecture 'Sur la Symetrie' at the French Mineralogical Society of Paris. Ten years later he concretized his ideas about symmetry in a paper (Curie 1894). Here is what he wrote:

The characteristic symmetry of a phenomenon is that symmetry which is best compatible with the existence of the phenomenon. A phenomenon can exist in surroundings which possess its characteristic symmetry or at least one subgroup of its characteristic symmetry. In other words, certain symmetry elements can exist together with certain phenomena but they are not necessary. But it is necessary that certain symmetry elements do not exist.

[For example, spontaneous polarization cannot exist in a crystal (which is the 'surroundings' here) that is centrosymmetric.]

Then he added the now classic statement: '*It is the dissymmetry that makes a phenomenon.*'

There has been considerable controversy and debate about what precisely did Curie mean in his various pronouncements. Even the validity of the principle has been questioned (see Ismael 1997). According to SK, '. . the obscure character and contradictory nature of a number of Curie's propositions have repeatedly provoked research workers to criticism and to the replacement of these propositions by other assertions based on the principle of causality or the principle of sufficient reason.'

There are two broad categories of the *apparent* violations of the Curie principle. One involves systems which interact so strongly (and therefore nonlinearly) with one another that a simple superposition of their individual (isolated) dissymmetries cannot give correctly the true interaction symmetry G_d (Ismael 1997; Nakamura and Nagahama 2000; Chiba and Nagahama 2001). Actually, this is more a case of not formulating the problem properly, rather than a violation of the principle.

The second category is what I discuss in this section, and it involves 'equal' or 'equivalent' component symmetries. This 'same-symmetry' superposition is best illustrated by a geometrical example.

Consider a parallelogram in two dimensions (Fig. 7.2(a)). It has only a 2-fold axis of symmetry.

Now make a mirror-image copy of it and construct the object in Fig. 7.2b. It has a mirror line of symmetry which is not in accordance with the Curie principle stated here so far. *This violation would not have occurred if the two parallelograms were not identical.*

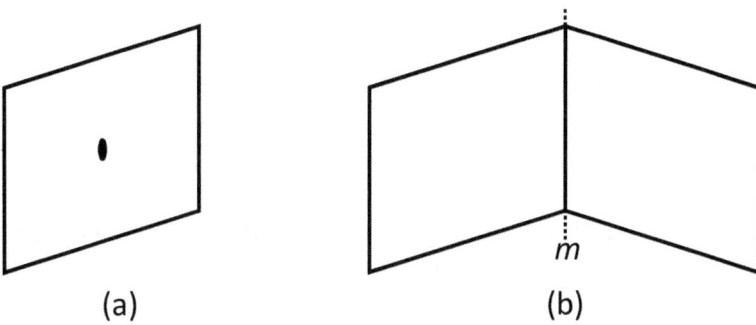

(a)   (b)

Fig. 7.2 (a) A parallelogram in two dimensions. The only symmetry it has is a 2-fold axis of rotational symmetry about its centre. (b) A composite object made by juxtaposing in a special way two such equal parallelograms. This object has symmetry higher than that predicted by Eq. 7.2. It has a mirror line of symmetry (*m*) not predicted by the original version of the Curie principle, a symmetry which has arisen because the two component parallelograms are of equal size, and are mutually placed in a special way.

Therefore, for the superposition or juxtaposition of equal or equivalent objects or phenomena the Curie principle must be generalized to account for the net symmetry, which is higher than G_d. The occurrence of this higher symmetry was called *symmetrization* by SK.

SK's book gives a generalized version of the Curie principle to accommodate the possibility of symmetrization. For Fig. 7.2b, $G_1 = 2$, $G_2 = 2$, and $G_d = 1$. According to SK, the net symmetry G_s of any composite object or phenomenon is given by the union of G_d and a set of *symmetrizers*. For Fig. 7.2(b),

$$G_s = G_d \cup (G_d \times m) = 1 \cup (1 \times m) = m. \tag{7.5}$$

Here m is a symmetrizer. This equation is an example of the generalized Curie principle, which I called *the Curie-Shubnikov principle* (Wadhawan 2000). The principle can be expressed as follows:

$$G_s = G_d \cup M, \tag{7.6}$$

where G_d is given by Eq. 7.2, and

$$M = G_d g_2 \cup G_d g_3 \cup ... G_d g_j \qquad\qquad (7.7)$$

Here $(g_2, g_3, ... g_j)$ are the representative elements of an appropriate system of cosets. For Fig. 7.2(b), only one symmetrizer $(g_2 = m)$ is needed.

Actually Curie (1894) did have an inkling of the symmetrization aspect. After all, he did say: '*Actions produced may be more symmetric than the causes.*'

Many examples of symmetrization are seen in Nature, the most familiar being that of crystals. Identical objects (namely the unit cells) are stacked together in three dimensions to constitute the composite object, namely the crystal ('*equal location of equal parts*' (Sheftal 1976)). The resulting symmetrization (translational symmetry) has a direct bearing on the diffraction pattern of the crystal, and on a host of other properties. Some other examples of symmetrization occur in: grain-boundary type bicrystals, chemical reactions, nuclear reactions, solutions of mathematical equations, external and internal conical refraction, etc. (Shubnikov and Koptsik 1974; Koptsik 1983).

Lastly, a word about naming the symmetry principle outlined here. I have stated above that Eq. 7.3 (the Neumann principle) follows from Eq. 7.2 (the Curie principle). But, as mentioned above, the converse is also true: We may well start from Eq. 7.3, and get Eq. 7.2. So both the principles are equally fundamental. In fact, they say the same thing. SK, as also Brandmuller (1986), state the symmetry principle as follows:

$$G_{object} \subseteq G_{property} \qquad\qquad (7.8)$$

And, for giving due credit to Minnigerode also, they call it the Neumann-Minnigerode-Curie (NMC) principle.

Eq. 7.8 does not cover the symmetrization aspect. We have seen above that this generalization was carried out by Shubnikov, and is embodied in Eq. 7.6 (to be read along with Eqs. 7.2 and 7.7). I had earlier called Eq. 7.6 the Curie-Shubnikov principle. A more equitable name would be the *Neumann-Minnigerode-Curie-Shubnikov principle*, or *the NMCS principle*.

7.3 Interplay between dissymmetrization and symmetrization

When a crystal grows from a fluid, the second law of thermodynamics for open systems permits (even ensures) a process of *symmetrization*: The assembly of the asymmetric units has higher symmetry than what is present in each asymmetric unit: There is at least the definitive translational symmetry of the crystal, and there may even be directional symmetry not present in each asymmetric unit. In fact, directional or point-group symmetry in crystals is very common.

The opposite process of *dissymmetrization* also occurs if we trace the process of crystal formation from atoms onwards (Table 7.1). Isolated atoms have the very high spherical symmetry. A molecule formed from the atoms has less symmetry than that of a sphere; it is one of the point-group symmetries, excluding spherical symmetry. So this is *lowering* of symmetry. And a large number of these molecules form a crystal, and then we get new symmetry elements not present in the molecules (a case of symmetry *enhancement*, or symmetrization).

Table 7.1 Successive decrease and increase of symmetry as we go from atoms to molecule to crystal.

Entity	Symmetry	Symmetry change
Atoms	Spherical	-
↓	↓	↓
Molecule	Less than spherical	Dissymmetrization
↓	↓	↓
Crystal	Crystallographic	Symmetrization

7.4 The Hermann theorem of crystal physics, and its applications

The Hermann theorem of crystal physics is stated as follows (Hermann 1934; also see Sirotin and Shaskolskaya 1982):

If a crystal possesses an N-fold axis of symmetry, and we consider a tensor property of rank r such that $r < N$, then this symmetry axis has the same effect on that tensor property as if $N = \infty$.

In other words, the crystal must exhibit complete isotropy (*transverse isotropy*) for that property in a plane perpendicular to the N-fold axis.

A familiar example of the validity of the theorem is provided by the nature of the optical indicatrix, describable by a polar tensor of rank $r = 2$. In agreement with this theorem, triclinic, monoclinic and orthorhombic crystal classes do not have optical isotropy normal to any symmetry axis of the crystal because none of the symmetry axes in them are of order higher than 2, meaning $N \leq r$. But crystal classes belonging to the trigonal, hexagonal, and tetragonal systems do have a symmetry axis of order greater than 2, so the Hermann theorem correctly predicts optical isotropy normal to it; this axis is the familiar *optic axis*. And there is only one optic axis for these crystal classes.

The cubic crystal classes are optically isotropic in all directions because there are more than one directions for which $N > 2$.

The theorem can be extended from crystals to composite materials, provided sufficient care is taken regarding the length scales involved. Laminated composites are particularly relevant in this context (Wadhawan 1987). The commonly used cross-ply plywood is an example of such a composite material.

Cross-ply plywood is a laminated composite, consisting of an odd number of plies of wood, bonded together so that the fibre axes of successive plies are at right angles to each other (Fig. 7.3). The two outermost plies are called the *faces*, and any interior plies having the same direction of grain or fibre axis as the faces are called *centres*. The plies with grain direction at 90^0 to that in the face plies are the *cores* (Countryman, Carney and Welsh 1969).

Thus, the design introduces symmetry about the central lamina, providing it a 'balanced structure'. The composite does not warp under changes of temperature and humidity, whereas any single ply would (Stavsky and Hoff 1969).

Such a structure can be viewed as made of two interpenetrating objects, one consisting of all the centres (symmetry $G_1 = mmm$), and the other consisting of all the cores (symmetry $G_2 = mmm$). The symmetrization operation for the composite can be taken as $g_2 = 4_z$. The point symmetry of cross-ply plywood is thus

$$G_s = (mmm) \cup 4_z(mmm) = 4/mmm \qquad (7.9)$$

In other words, this composite is so designed that a pseudo-4-fold axis of symmetry exists normal to its principal plane; thus $N = 4$, preventing warping and anisotropic transverse thermal expansion or contraction: The composite has transverse isotropy for these properties because they are rank-2 tensor properties.

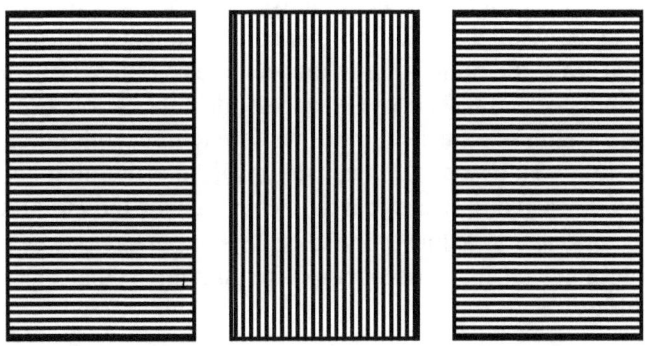

Fig. 7.3 The structure of cross-ply plywood. There is an odd number of plies (3, 5, 7, ..). The grains in the successive plies are oriented in the sequence shown in this figure. With respect to a horizontal x-axis, the grains in the successive plies make angles of $0°$, $90°$, $0°$, $90°$, $0°$, ..

7.5 Hexply configurations for nanocomposites

Suppose we consider the *elastic stiffness tensor* of a cross-ply composite laminate. Transverse isotropy does not exist for this tensor property because $r = 4$ for it and thus $N = r$, rather than $N > r$. This is borne out by detailed calculation. The group $4/mmm$ allows 6 independent nonzero components for this tensor property, whereas only 5 would have been allowed had the effective symmetry for this property been ∞/mm.

To achieve the effective symmetry ∞/mm for this rank-4 tensor property, we must design the plywood such that $N \geq 5$. The Hermann theorem says that transverse isotropy can then be achieved for any rank-4 property of the composite. Configurations with $N = 6$ and $N = 8$ are some of the practically convenient possibilities.

The configuration with $N = 6$ (the so-called *hexply*) would involve repeated stacking and bonding of a basic unit consisting of three plies such that their grain directions make angles of $0°$, $60°$, and $120°$ with some arbitrary direction in the principal plane. The resultant configuration has the point-group symmetry 622. Application of the Hermann theorem gives the result that for a tensor property of rank 4, such a configuration would exhibit the point symmetry $\infty 2$, implying transverse isotropy of elastic behaviour (Wadhawan 1987).

Similar considerations can be applied to the *photoelastic tensor* also, another rank-4 property. A variety of nanocomposites are being designed for optical applications (see Wadhawan 2007). In a nanocomposite, at least one of the phases has at least one dimension in the nanometre range, i.e., much below the wavelength of light. The easily accessible fabrication techniques for making flat or laminated nanocomposites (stretching, rolling, or poling of thin layers, followed by bonding) usually result in orthorhombic symmetry (meaning $N = 2$) and optically biaxial behaviour for the individual layers. It should be possible to achieve optically uniaxial behaviour by adopting a $4/mmm$ configuration for a nanocomposite, similar to that of the familiar cross-ply plywood.

But if it is desired that the nanocomposite should be not only optically uniaxial, but also possess transverse isotropy of photoelastic behaviour (rank-4 tensor property), then the hexply configuration is necessary. *Such a material would be free from vibration-induced fluctuations of transverse birefringence* (Wadhawan, unpublished).

8. Gauge Symmetry

Gauge symmetries are formulated only in terms of the laws of nature; the application of the symmetry transformation merely changes our description of the same physical situation, (it) does not lead to a different physical situation.

David Gross (1996)

Indeed today we believe that global symmetries are unnatural. They smell of action at a distance. We now suspect that all fundamental symmetries are local gauge symmetries. Global symmetries are either all broken (such as parity, time reversal invariance, and charge symmetry) or approximate (such as isotopic spin invariance) or they are the remnants of spontaneously broken local symmetries.

David Gross (1996)

(Gauge symmetry) is so important that essentially all of modern physical theory is based on it. But it is so subtle that without using mathematics, it is hard to describe. It is so subtle that its ramifications are still being unravelled today, more than a hundred years since it was first suggested.

Lawrence Krauss (2017)

8.1 Introduction

We have implicitly assumed so far in this book that the transformations under which questions about the presence or absence of symmetry are asked are *global transformations*. We have chosen a globally applicable reference frame throughout. It is remarkable that Nature reveals some fundamentally important symmetries (called gauge symmetries) under certain *local transformations*. This may sound counter-intuitive, even meaningless, but it is not.

Imagine a network of local reference frames. It is found sometimes that certain transformations which are not the same for all of them (they differ from location to location, or from one instant to another) can still be symmetry transformations. In such *local* transformations there is at least one space-time-dependent parameter. *Global symmetry can then be viewed as a special case of this, wherein the local transformations are the same throughout, and there is no space and/or time dependent parameter.*

'Gauge' means 'measure'. It is synonymous with 'reference frame'. The gauge used for measuring tyre pressure is an example. Its calibration provides the reference frame for measuring pressure. A global reference gauge for tyre pressure would mean that all the gauges are calibrated identically. By contrast, local reference gauges would correspond to different calibrations at different locations (or times).

Gauge symmetry means invariance under local change of measure. The term *gauge transformation* usually connotes *local* transformations (recalibrations) of the reference frames.

The forces or interactions among elementary particles are described in terms of *fields*. In the *field theories* of the interactions, the fundamental fields cannot be measured directly. Instead, some associated parameters (observables) such as charge, energy, or velocity have to be measured. There can be a possible degeneracy involved in this scenario, in that different configurations of the (unobservable) fields may correspond to the same observables. Gauge transformations take us from one field configuration to another. The corresponding observables are said to be *gauge invariant*.

Let us look at some examples.

1. Consider a ball of mass m on a staircase at a height h_1 from some baseline. Its gravitational potential energy is mgh_1. If it falls by a step on the staircase to a height h_2, the new potential energy is mgh_2. The difference ($mgh_1 - mgh_2$) is invariant ('gauge invariant') to the choice of the baseline for measuring the height h. Einstein's theory of gravitation is a *gauge-invariant theory* because its predictions remain the same when the baseline is changed. We say that it is a field theory that exhibits *gauge symmetry*.

2. Another example of gauge invariance comes from electromagnetism. The magnetic vector potential **A** (a vector field) is defined to be such that its curl is the magnetic field **B**. If a scalar field φ is introduced similarly for the electric field **E**, the magnetic and electric field can be defined as follows:

$$\mathbf{B} = \nabla \times \mathbf{A} \tag{8.1}$$

$$\mathbf{E} = -\nabla\varphi - \frac{\partial \mathbf{A}}{\partial t} \tag{8.2}$$

This definition of magnetic vector potential **A** is not unique because of the gauge invariance involved: One can shift the base line of the scale by adding curl-free components to the vector potential without affecting the observed magnetic field.

The electric field **E** and the magnetic field **B** are *observables*. The vector potential **A** is not an observable. A gauge transformation in which a constant term is added to **A** makes no change to the observable fields **B** and **E**. Similarly, **E**, being the gradient of a scalar (electric potential) is not affected if a constant term is added to the electric potential.

[A counter example is that of relative speed (because of the effects of special relativity). Suppose two objects pass an observer, one at 9% the speed of light, and the other at 20%. The observer will measure a speed difference of 11%. But if the observer is travelling at 50% of the speed of light, the observed difference in the speeds of the two objects will be only 7% of the speed of light. Gauge symmetry is absent in this case because there is lack of invariance under a change of the reference point.]

3. Our universe is charge-neutral. An electron has a negative charge and a proton has a positive charge, but this labelling is arbitrary. If we reversed the signs of charge of all the fundamental particles, the laws of physics would still be the same, and the universe would still be the same. This invariance under reversal of charge implies a symmetry, the *charge-reversal symmetry*. And this symmetry implies the *conservation of charge*: The net charge is conserved: It is not possible for any charge to appear spontaneously, without an equal and opposite charge appearing simultaneously (otherwise the charge-reversal symmetry would be violated).

Krauss (2017) has given the analogy of a chessboard to explain how the law of conservation of charge follows from the charge-reversal symmetry of the universe. The analogy is further used for introducing the idea of gauge invariance. If all the white squares on a chessboard are changed to black, and all black to white, the game of chess can be played as before after rotating the board by 180°. This is global symmetry. Next, suppose white is changed to black for some square, but black is not changed to white for a neighbouring square. This is a *local* (rather than global) transformation, resulting in two neighbouring black squares. Now the game cannot be played as before, *with the existing rules book*. But we can have a new rules book, in which the rules are specified for each game piece when two adjacent black squares are encountered. The game can then be played as if no local transformation ('gauge transformation') has been made, although this comes with some crucial consequences. To understand that, let us not stretch the chessboard analogy too much, and return to the charge-conservation case.

First a little bit of jargon, namely the formal meanings of 'function' and 'field'. In set theory, a point field is a (finite or infinite) set of elements called *points* (an example being the set of squares on a chessboard). A mapping from a point field *A* to a point field *B* is said to be defined if, for every point *p* in *A*, a point *p'* is associated in *B*. Such a mapping defines a *function f*. A function associates a rule with each point of the point field.

In physics, a function defined at every point in space is called a *field*. An example is the electromagnetic field. It describes the strength of electric and magnetic forces at each point in space.

For the charge-conservation example, it turns out that the properties of the function that allow us to change our definition of electric charge from point to point, without changing the underlying physics determining the interaction among the charges, are the same as those that determine the form of the rules that an electromagnetic field must obey (namely, the Maxwell equations). Thus, the requirement that a local gauge transformation (namely a local change in the assignment of signs of charges) be a symmetry transformation requires the existence of an electromagnetic field that obeys Maxwell's equations. This *gauge invariance* entirely determines the nature of electromagnetism. To quote Krauss (2017): 'Gauge symmetry in electromagnetism says that I can actually change my definition of what a positive charge is locally at each point of space without changing the fundamental laws associated with electric charge, so long as I also somehow introduce some quantity that helps keep track of this change of definition from point to point. This quantity turns out to be the electromagnetic field.'

8.2 Gauge-symmetry groups

That gauge symmetry has such a strange name has little to do with quantum electrodynamics and is an anachronism, related to a property of Einstein's General Theory of Relativity, which, like all other fundamental theories, also possesses gauge symmetry. Einstein showed that we are free to choose any local coordinate system we want to describe the space around us, but the function, or field, that tells us how to connect these coordinate systems from point to point is related to the underlying curvature of space, determined by the energy and momentum of space. The coupling of this field, which we recognize as the gravitational field, to matter, is precisely determined by the invariance of the geometry of space under the choice of different coordinate systems.

Lawrence Krauss (2017)

There used to be a minority viewpoint that gauge symmetry is no symmetry at all (Carroll 2005). But now everybody agrees that introduction and application of the concept of gauge symmetry has resulted is several remarkable advances in fundamental physics.

Global passive transformations can constitute a group. For example, all rotations about a point constitute a 3-parameter group, the three parameters being, say, the three Euler angles. Another example is the set of all displacements (x, y, z) along the three coordinate axes. A third example of such a 3-parameter group is the set of all 'boosts' (velocity changes, $\Delta v_x, \Delta v_y, \Delta v_z$).

Such global transformations can be made *local* (or 'gauged') by making them space and/or time dependent. The resulting 'gauge transformations' can also constitute a group, the *gauge group*. ['Global-symmetry-derived gauge group' would be a more appropriate expression.]

When such gauge transformations are carried out, the system may behave as if there are *forces* present which are effecting the rotations or the changes of velocity etc. The 6-parameter *Lorentz group* provides a good example of that (three parameters are needed to specify any position, and three more for any velocity-boosts up to the speed of light). Just like the constant-velocity evolution of a massive particle, a Lorentz-transformation-effected evolution should also be possible. But the latter implies possible *acceleration*, which in turn means that a force must be at play, implying that *the notion of a 'free' massive particle is a myth*. This is how gravitational field enters the general theory of relativity, which is a gauge theory. It is the simplest nontrivial theory that possesses gauge symmetry under Lorentz gauge transformations (Rosen 2008).

Gauge theories play a central role in the development of fundamental theories of physical laws. All the fundamental interactions (the electroweak interaction, the strong interaction, and the gravitational interaction) are described by Lagrangians possessing gauge invariance.

Several fundamental laws in particle physics are also governed by gauge symmetry. The gauge-symmetry group for the electromagnetic interaction is

$U(1)$. And that for the weak nuclear interaction is $SU(2)$. Turning things upside down, we can say that *it is global symmetry which is a special case of gauge symmetry, and not the other way round.*

Gauge symmetry imposes severe restrictions on the dynamics of a system. For example, the global symmetry of the strong nuclear interaction is described by the 8-parameter 'colour' transformation group $SU(3)$, its operations acting on the colour states of the quarks (constituents of nucleons and pions). It was realized that since a constant-velocity proton is a possible inertial evolution, so should be any noninertial, *identity-changing*, evolution obtained from it by the action of a colour gauge transformation. This consideration pointed to the incompleteness of the original picture comprising solely of a proton, and required the postulation of additional entities (*gluons*) interacting with the quarks inside the proton. This is how in the field theory called quantum chromodynamics (QCD), the picture of quarks emitting and absorbing gluons for interacting with other quarks was built up.

In physics, a gauge theory is a type of field theory in which the Lagrangian is invariant under a *continuous group of local transformations*. The celebrated Noether's theorems lie at the heart of gauge theory, with fundamental implications for particle physics in particular, and symmetry and physics in general.

8.3 Noether's theorems

First a word about the remarkable Emmy Noether. I quote Sweatman (2016): 'Einstein referred to her as the most important woman in the history of mathematics. Her theorem has been recognized as *"one of the most important mathematical theorems ever proved in guiding the development of modern physics."* Yet many people haven't the slightest clue of who this woman was, or what she did that was so significant to our understanding of how our world works. If you count yourself as one of those who have never heard of Emmy Noether and wish to enlighten yourself, please read on. I can only hope I do her memory justice. Not just by telling you who she was, but by also giving you an understanding of how her insight led to the coming together of symmetry and quantum theory, pointing academia's arrow toward quantum electrodynamics.

'Being a female in Germany in the late 1800s was not easy. She wasn't allowed to register for math classes. Fortunately, her father happened to be a math professor, which allowed her to sit in on many of his classes. She took one of his final exams in 1904 and did so well that she was granted a bachelor's degree. This allowed her to "officially" register in a math graduate program. Three years later, she earned one of the first PhD's given to a woman in Germany. She was just 25 years old.

'1907 was a very exciting time in theoretical physics, as scientists were hot on the heels of figuring out how light and atoms interact with each other. Emmy wanted in on the fun, but being a woman made this difficult. She wasn't allowed to hold a teaching position, so she worked as an unpaid assistant, surviving on a small inheritance and under-the-table money that she earned sitting in for male professors when they were unable to teach. She was still able to do what professors are supposed to do, however – write papers. In 1916, she would pen the theorem that would have her rubbing shoulders with the other physics and mathematical giants of the era.'

There is a deep and universal connection between symmetry and the conservation laws of physics. This realisation dawned and grew rather gradually in the history of physics. *The deep and unexpected relationship between conservation laws and the symmetries of nature has been the single most important guiding principle in physics in the past century* (Krauss 2017). And Noether's two theorems are formal and precise statements of this relationship. They are crucial for understanding the fabric of our reality (Browne 2011).

Many, if not all, laws of fundamental physics can be understood and formulated entirely from symmetry considerations. As stated repeatedly in this book, symmetry means invariance or conservation of something under certain operations.

There is an interesting and important background that led to the two theorems proved by Noether. I quote Browne (2011): 'Proven in 1915 and published in 1918, Noether's theorem, as applied to physics, tells us about conservation laws and conserved quantities in dynamical systems. In 1915, general relativity was almost a finished theory, but there was a problem regarding the conservation of energy. Hilbert had noted that the failure of general relativity to produce a classical conservation of energy equation was intrinsic to the system. He asked Noether to help clarify his idea. Noether proceeded to derive two foundational theorems which confirmed Hilbert's suspicion and fundamentally changed the way that the Lagrangian formalism was seen.'

Noether (1918) proved two different theorems in her famous paper. The first deals with *global* symmetries (generated by *finite* Lie groups) and states that these symmetries lead to conserved 'charges'.

The second theorem is about gauge symmetries. They are of a *local* nature in general (governed by *infinite-dimensional* Lie groups), involving arbitrary functions of spacetime (Einstein's theory of gravity is an example). The second theorem states that the gauge symmetries inevitably lead to relations among the equations of motion.

Noether's first theorem

Noether's first theorem applies to any system that can be derived from an 'action' (see below) and possesses some continuous, non-gauge, symmetry: For any given symmetry, Noether's algorithm associates a conserved 'charge' to it. It can be stated roughly as follows: *For every continuous symmetry of the laws of physics, there must exist a corresponding conserved dynamical variable or quantity; and for every conservation law of such a variable or quantity there must exist a continuous symmetry.*

A somewhat more precise statement of Noether's first theorem is: *Every differentiable symmetry of the action of a physical system has a corresponding conservation law.*

The terms 'differentiable symmetry' and 'action' may not be widely familiar, so I explain them here.

The *symmetric derivative* is a generalization of the usual, 'ordinary' derivative, and is defined at a point x as $lim_{h \to 0} \frac{f(x+h)-f(x-h)}{2h}$. It is the limiting value of the 'symmetric difference quotient' around that point, or the arithmetic mean of the left derivative and the right derivative at that point, if the latter two exist. A function possesses differentiable or differential symmetry at a point x if its symmetric derivative at that point exists.

If a function is differentiable at a point, then it is also symmetrically differentiable at that point. But the converse is not true, an example being the function $f(x) = |x|$. Because of the discontinuity at $x = 0$, it is not differentiable (i.e., its ordinary derivative does not exist) at that point, but is *symmetrically* differentiable there, the symmetric derivative being 0.

And now about 'action'. The crucial concept exploited by Noether is that of an 'action symmetry'.

Any dynamical system evolves with time. It has a certain kinetic energy and potential energy at each instant. The *action function* at time t is defined as the kinetic energy *minus* the potential energy at that time, and the all-important *principle of least action* says that a dynamical system takes that trajectory between any two times t_1 and t_2 for which the time-integral of the action between these two points, namely the Lagrangian $\mathcal{L}$, has the least possible value. In other words, the action must be stationary under arbitrary variations of the dynamical variables.

In classification mechanics, the action I is defined by

$$I[q^i(t)] = \int dt\, \mathcal{L}(q^i, \dot{q}^i, t) \tag{8.3}$$

There are an associated set of Lagrange equations, derived from an extremum principle with fixed points:

$$\frac{d}{dt}\left(\frac{\partial \mathcal{L}}{\partial \dot{q}^i}\right) - \frac{\partial \mathcal{L}}{\partial q^i} = 0. \tag{8.4}$$

The principle of least action says that

$$\delta I[q^i(t)] = 0. \tag{8.5}$$

The variables $q^i(t)$ can be the locations of particles in space, fluctuations of fields, or auxiliary fields.

A mathematical equation has been derived by Banados and Reyes (2017) that embodies the conservation law stated in Noether's first theorem:

$$\frac{d}{dt} Q = 0 \quad \text{with} \quad Q = K - \frac{\partial \mathcal{L}}{\partial \dot{q}^i} \delta_s q^i. \tag{8.6}$$

In words: Given an action symmetry $\delta_s q^i(t)$, the combination Q defined above is conserved. [K in Eq. 8.6 is a boundary term. As discussed by Banados and Reyes (2017), K does not interfere with the *existence* of a conserved charge; and for many important examples it is non-zero and *contributes* to the charge.]

Here is another statement of Noether's first theorem: *To every differentiable symmetry generated by local actions, there corresponds a conserved current.*

Finally, here is a formal statement of Noether's first theorem (Brown 2011):

Given a Lie group G, whose most general transform depends on ρ parameters, under the action of which an integral I is invariant, there are ρ linearly independent combinations of the Lagrange expressions which become divergences. The converse also holds true; the existence of ρ Lagrange expressions which are divergences implies invariance of I under the action of a transform in G.

The Lagrange expressions mentioned above are the left-hand side of the familiar Euler-Lagrange equations (Eq. 8.4), namely $\frac{d\mathcal{L}}{dx_i} - \frac{d}{dx_i}\left(\frac{d\mathcal{L}}{d\dot{x}_i}\right)$. When a system obeys Hamilton's principle, the value of these expressions is 0 (Eq. 8.4).

For a rigorous discussion of Noether's theorems, see Browne (2011), as also the entry in Wikipedia under heading 'Noether's theorem' (https://en.wikipedia.org/wiki/Noether%27s_theorem). I quote from the Wikipedia to introduce some more terminology:

'The word "symmetry" [in the statement of the Noether theorem] refers more precisely to the covariance of the form that a physical law takes with respect to a one-dimensional Lie group of transformations satisfying certain technical criteria. The conservation law of a physical quantity is usually expressed as a continuity equation.

'The formal proof of the theorem utilizes the condition of invariance to derive an expression for a current associated with a conserved physical quantity. In modern (since *ca.* 1980) terminology, the conserved quantity is called the *Noether charge*, while the flow carrying that charge is called the *Noether current*. The Noether current is defined up to a solenoidal (divergenceless) vector field.'

Examples of the applicability of Noether's first theorem abound.

1. Laws of physics are the same for all instants time. So, there is time-symmetry. The corresponding conserved quantity is the total energy of the universe. That the total energy can be neither increased nor decreased is nothing but a statement of *the law of conservation of energy*, or the first law of thermodynamics.

2. Laws of physics are the same for all points of space. This is translational

symmetry. The corresponding conserved quantity demanded by Noether's first theorem is the total linear momentum. Newton's first law of motion is nothing but a statement of that fact.

3. Space has isotropy symmetry also: The laws of Nature are the same in all directions. There is complete rotational symmetry This is the reason behind the law of conservation of the total angular momentum.

Noether's first theorem discussed above cannot be applied for any gauge theory. Her second theorem must be invoked for gauge theories. The first theorem applies only to systems which undergo *global* transformations, whereas gauge theories require the Lagrangian to be invariant under *local* gauge transformations.

Noether's second theorem

Noether's second theorem states that *if the action function has an infinite-dimensional Lie algebra of infinitesimal symmetries parameterized linearly by k arbitrary functions and their derivatives up to order m, then the functional derivatives of the Lagrangian satisfy a system of k differential equations.*

The conservation laws derived from Noether's second theorem are dependent on the metric (or shape) of the space that is being analysed, and a conservation law derived using one set of coordinates cannot be used in another set of coordinates. In this scenario, a law of conservation of momentum, derived in cartesian coordinates, would not be applicable to a system described by polar coordinates (Browne 2011).

The second theorem is less widely known (or its existence is just ignored for reasons of context or convenience), and the first theorem is often referred to as *the* Noether theorem. But we should not lose sight of the fact that the first theorem is about *global* symmetries, whereas gauge symmetries (which occupy centre stage in modern physics) are the focus of the second theorem. The current view point is that gauge symmetries are the real thing, and that global symmetries are only the limiting cases of gauge symmetries.

9. Phase Transitions and Broken Symmetry

What is the common link between symmetry and complexity? It is symmetry breaking as the origin of dynamics and variety of forms and systems in the world. Thus, symmetry and complexity are the spirit of nonlinear science.

Klaus Mainzer, *Symmetry and Complexity*

When a physical system does not exhibit the full symmetry of the laws governing it, we speak of broken symmetry.

Why does symmetry break spontaneously under certain conditions? I answer this question in this chapter, and then discuss some symmetry aspects of phase transitions in crystals and some other systems. A phase transition usually (but not always) involves a change of symmetry. Landau's (1937) work heralded the era in which we pay detailed attention to the changes of symmetry across a phase transition. He introduced the all-important notion of an '*order parameter*' emerging at a phase transition and breaking the symmetry of the parent phase.

The scope of broken-symmetry investigations goes far beyond condensed-matter physics (see, e.g., Zee and Penrose 2016). In particle physics, search for new broken symmetries is a mainstream activity. We shall see in this and the next chapter why this is so. Even in biology, studies on broken symmetry provide deep insights into the very nature of Darwinian evolution. This also will be discussed.

9.1 Liberal meanings of the term 'phase transition'

In the context of complex systems (Wadhawan 2010, 2017), the term 'phase transition' tends to be used in a more liberal manner than what it means when inanimate matter in a *simple* system makes a transition to a different, lower-free-energy phase under the action of some control parameter. For example, the transition from chaos to order at the 'edge of chaos' is viewed as a 'phase transition' from a chaotic state to a state of self-organized order and higher degree of complexity.

Here is an example of an even more liberal definition of a phase transition, taken from recent literature on complexity (Achlioptas, D'Souza and

Spencer 2009):

'A large system is said to undergo a phase transition when one or more of its properties change abruptly after a slight change in a controlling variable. Besides water turning into ice or steam, other prototypical phase transitions are the spontaneous emergence of magnetization and superconductivity in metals, the epidemic spread of disease, and the dramatic change in connectivity of networks and lattices known as percolation.'

9.2 Spontaneous breaking of symmetry

Congratulations. I knew the record would stand until it was broken.
                                                                Yogi Berra

As an example of broken symmetry, consider a crystal of iron cooled through the 'critical temperature' T_c. It undergoes a transition from the paramagnetic phase to the ferromagnetic phase. In the ferromagnetic phase there is a spontaneous magnetization and there is a preferred direction along which the spontaneous magnetization points, whereas there was no such preferred direction in the paramagnetic phase. Why did this phase transition occur? It occurred because the crystal can lower its free energy on changing over to the ferromagnetic phase.

But why is there a spontaneous lowering (breaking) of symmetry when all we have done is to cool the crystal? Why has a lower-symmetry (ferromagnetic) phase arisen? According to the Curie principle of symmetry discussed in Section 7.1, the effect cannot be less symmetric than the cause. The cause (temperature) in this case is apparently an isotropic scalar field, with very high symmetry. The effect (the ferromagnetic phase) should not have a lower symmetry than the symmetry of the paramagnetic phase. Has the symmetry principle been violated? No.

'Temperature' is a number indicative of the *average* kinetic energy of a large number of molecules. It is a macroscopic concept, and the underlying phenomena are of a statistical nature. At the microscopic level there are *thermal fluctuations* of the crystal structure. And each such fluctuation is a symmetry-breaking spontaneous perturbation. Any of the anisotropic fluctuations (in an *unstable equilibrium* situation) could have caused the lowering of symmetry at the temperature T_c, rather like the situation depicted in Fig. 9.1.

The higher-symmetry (paramagnetic) phase is unstable at T_c. This unstable-equilibrium situation is schematically represented by the ball perched at the top of the potential-energy surface in Fig. 9.1. Even the tiniest of perturbations (thermal fluctuations) can send the ball rolling to the rim of the Mexican-hat potential, with a concomitant breaking of symmetry. Thus, there is no violation of the Curie principle if we look minutely enough.

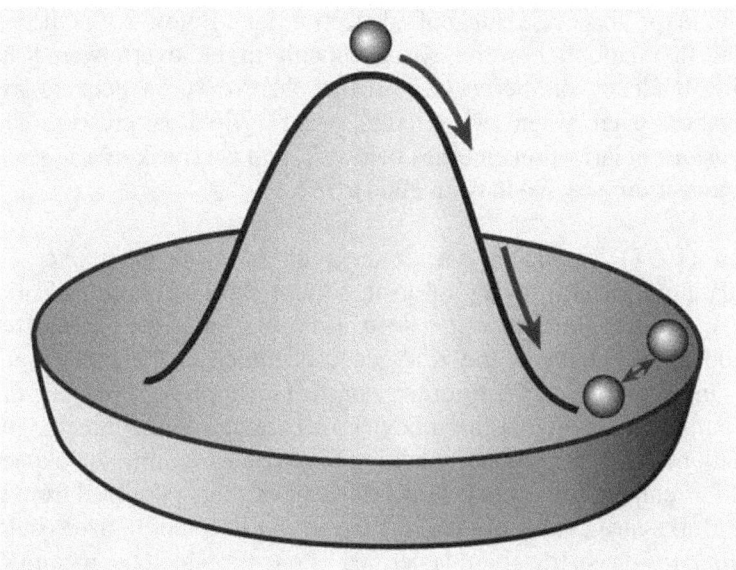

Fig. 9.1 An illustration of spontaneous breaking of symmetry. The ball could have rolled down along any direction, but once a random choice has been made spontaneously, the final state has a lower ('broken') symmetry. This figure also illustrates 'unstable equilibrium'. The top position has zero slope, and therefore corresponds to an equilibrium. But this is *unstable* equilibrium because it is not robust against even a minor displacement away from equilibrium. Unstable equilibrium leads to spontaneous breaking of symmetry.
Image credit: http://cerncourier.com/cws/article/cern/32522/1/CChig5_01_08

9.3 The Landau theory of phase transitions

Cosmic evolution leads from symmetry to complexity by symmetry breaking and phase transitions. The emergence of new order and structure is explained by physical, chemical, biological and social self-organization, according to the laws of nonlinear dynamics. All these dynamical systems are considered computational systems processing information and entropy.

Klaus Mainzer, *Symmetry and Complexity*

The free energy of a crystal changes as we vary the control parameter, usually temperature. At and below a certain critical temperature T_c a different competing phase of the crystal has a lower free energy than the phase the crystal is in. It therefore makes a *transition* to the new phase. For example, suppose that we are dealing with a large crystal, and above T_c the crystal is in a paraelectric phase, characterized by a zero average dipole moment in the absence of an applied electric field. Below T_c the disordering thermal fluctuations become weak enough to be overpowered by the ordering tendency of the crystal, so that there exists a nonzero average polarization even when no external electric field is present. Thus, a *spontaneous* polarization emerges below T_c, and we speak of a *ferroelectric* phase transition (see Wadhawan 2000).

Landau (1937) introduced the concept of an order parameter (η) for formulating a general theory of an important class of phase transitions in crystals, namely *continuous phase transitions*. The order parameter is a thermodynamic quantity, the emergence of which at the phase transition results in a lowering of the symmetry of the parent phase of the crystal. The order parameter, like other thermodynamic parameters, is subject to thermal fluctuations. It is envisaged as having a zero mean value, or expectation value, for temperatures above T_c. As the temperature is lowered from values above T_c, the mean value of the order parameter becomes nonzero below T_c. A *continuous* phase transition is defined as one for which the mean value of the order parameter rises *continuously* (rather than abruptly) from the value zero as the temperature is reduced from the value T_c, i.e. as the system enters the daughter phase. [We assume here that the phase transition to the ordered phase occurs on cooling, rather than on heating; this is usually the case.]

The emergence of the order parameter breaks the symmetry of the parent phase, in accordance with the Curie principle. As an illustration, consider the ferroelectric phase transition that occurs in a crystal of $BaTiO_3$ as it is cooled below 130°C at atmospheric pressure. Above this temperature (T_c) the crystal symmetry is cubic (Fig. 9.2). The order parameter (electric polarization) has the symmetry of a one-headed arrow or a cone, and it is directed along one of the edges of the cubic unit cell of the parent phase.

The Curie principle demands that the daughter phase have only those, and all those, symmetry elements that are common to the symmetry of the cube and the symmetry of the order parameter (or that of an arrow pointing along an edge of the cube). An arrow has infinite-fold rotational symmetry along its length, and the cube has only a 4-fold symmetry along an edge direction. So the lower of the two, namely the 4-fold axis of symmetry, survives the superposition of the two symmetries. In addition, any plane passing through the length of the arrow is a mirror plane of symmetry, so the arrow has an infinity of such symmetry elements, m. But the cube has only two such mirror-symmetry planes, namely those normal to the x-direction and the xy-direction (we have taken the direction of the arrow as the z-direction). So only these two mirror planes are common between the cube and the arrow, and survive across the phase transition.

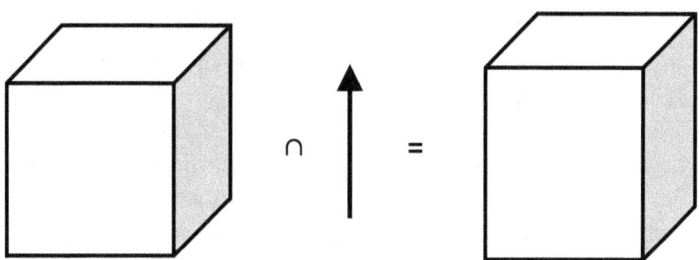

Fig. 9.2 Dissymmetrization occurs across the cubic-to-tetragonal phase transition in BaTiO$_3$.

Thus the net surviving symmetry after the phase transition is $4_z m_x m_{xy}$, commonly written as just $4mm$. This is the symmetry of a square prism (shown on the right-most part of Fig. 9.2). This is how the emergence of a lower symmetry across a continuous phase transition is explained by the Landau theory and the Curie principle.

9.4 Ferroic phase transitions and domain structure

Ferroelectric phase transitions, characterized by the emergence of spontaneous polarization (which is a macroscopic tensor property), are an example of a more general class called *ferroic phase transitions* (FPTs) (Wadhawan 2000).

Let us consider phase transitions involving a change of the space-group symmetry of a crystal. Space-group symmetry has a translational-symmetry part and a directional or point-group-symmetry part. The translational part is described by one of the 14 Bravais groups, and the directional part by one of the 32 crystallographic point groups. If there is a change of the point-group symmetry, then we have, by definition, a ferroic phase transition, or FPT. If only the translational part of the symmetry changes and the point-group symmetry remains the same, we speak of a nonferroic phase transition. A *ferroic material* is one which can undergo at least one FPT (Wadhawan 2000).

Across an FPT, the parent phase has a higher point-group symmetry than the daughter phase or ferroic phase. The orders of the two point groups differ by at least a factor of 2 (because of the Lagrange theorem for subgroups; see Section 3.3). This means that the ferroic phase must have *domains* (or *variants*). For example, if there is a ferromagnetic phase transition, one type of domains will have spin up, and the other type spin down. Similarly, in the case of the ferroelectric phase transition in $BaTiO_3$ at $T_c = 130°C$, six ferroelectric domain-types or variants can arise. This is because the point-group symmetry above T_c is cubic ($m\bar{3}m$) (with no preferred or 'polar' direction), and the point-group symmetry in the ferroelectric phase below T_c is 4*mm*, which is a polar group belonging to the tetragonal crystal system. The 4-fold axis is the polar axis, and it can point along any of six possible directions: $+x$, $-x$, $+y$, $-y$, $+z$, $-z$. Each of these possibilities gives rise to a distinct ferroelectric *domain type*. This is why we get domain structure in the ferroic phase.

9.5 Prototype symmetry

The concept of *prototype symmetry* is crucial for a proper description of FPTs. A crystal may undergo a sequence of symmetry-lowering phase transitions, say on cooling. Any phase in this sequence has a parent phase, from which it arose on cooling. Prototype symmetry is not just the symmetry of the next higher 'parent' phase. It is the *highest* symmetry conceivable for that crystal structure, so that all the daughter phases can be considered as derived from it by the loss of one or more symmetry operators. For example, $BaTiO_3$ undergoes a sequence of FPTs on cooling: From cubic to tetragonal to orthorhombic to rhombohedral. For each of the lower-symmetry phases, it is the cubic-phase symmetry which is the prototypic symmetry. For a rigorous definition of prototype symmetry, see Wadhawan (1998, 2000).

9.6 The symmetry compensation law

If one looks into the causes of spontaneous breaking of symmetry, the following symmetry compensation law appears to hold (Shubnikov and Koptsik 1974):

If symmetry is reduced at one structural level, it arises and is preserved at another.

The occurrence of domain structure in a ferroic crystal is an example of this. In Section 9.2 above I argued how, although temperature is apparently a scalar field, at the microscopic level the thermal fluctuation which happens to be present when the crystal goes over to the ferromagnetic phase decides the orientation of the spontaneous magnetization. But different portions of the specimen crystal may undergo different thermal fluctuations at the moment of the phase transition. Therefore, there is no reason why the whole specimen should prefer only one orientation of the spontaneous magnetization. All allowed orientations are equally likely to occur. That is exactly why there is a domain structure, with different domains having different allowed directions for the spin.

That is not all. One can map one domain type to another by performing the same coordinate transformations which were *symmetry* transformations before the spontaneous breaking of symmetry occurred. In other words, the domain structure of the ferroic phase carries the full signature of the symmetry broken across the phase transition. Thus there is symmetry compensation for the crystal specimen as a whole.

This can be illustrated by considering the example of the cubic-to-tetragonal ferroelectric phase transition in $BaTiO_3$ considered above (cf. Fig. 9.2). There are six domain types in the ferroelectric phase, corresponding to the six directions along which the spontaneous-polarization vector can point, namely $+x$, $-x$, $+y$, $-y$, $+z$, $-z$. A number of symmetry elements have been lost across this phase transition. One is the inversion operation, which makes any point (x, y, z) equivalent to the point $(-x, -y, -z)$. This is no longer a symmetry operation in the noncentrosymmetric ferroelectric phase. But it becomes a mapping operation in the ferroelectric phase, taking, e.g., the domain type $+x$ to the domain type $-x$.

9.7 Continuous broken symmetries

> *The more theoretical physicists penetrate the ultimate secrets of the microscopic nature of the universe, the more the grand design seems to be ultimate symmetry and ultimate simplicity. But all of the interesting parts of the universe, at least to us, are, like the earth itself as well as our own bodies, markedly complex and markedly unsymmetric. In the most elementary sense, then, we are surrounded by "broken symmetry", the result undoubtedly of some sequence of catastrophes.*
>
> P. W. Anderson

A symmetry broken spontaneously may be either continuous or discrete. In particle physics, the idea of continuous broken symmetries and of symmetry-restoring fluctuations (now called *Nambu-Goldstone modes*) was first put forward by Goldstone (1961). In condensed-matter physics, discussion of such ideas can be traced to the much earlier work of Landau, done in the 1930s. A major boost to the notion that broken symmetry occupies centre stage in the physics of condensed mater came for the work of Anderson (1972, 1981, 1984).

Consider a system having symmetry described by a continuous group G_0. Naturally, its Lagrangian density L is invariant under the operations of G_0. The system may have either a single nondegenerate ground state, or more than one degenerate ground states. An example of the former case is the s-state of a free atom. It is invariant under the operations of G_0, which in this case represents full spherical symmetry.

But if the system possesses degenerate ground states (i.e., different eigenstates, but same eigenvalue), then it is possible that some operation(s) of G_0 map one such state to *another*, rather than leaving it invariant. We then speak of broken symmetry, because the ground state is no longer invariant under the operations of G_0.

Fig. 9.1 illustrates the point, although in a somewhat different context. The various positions of the ball in the brim of the Mexican-hat potential correspond to the different degenerate ground states, and the symmetry operations of the cylindrical symmetry group of the potential take the ball from one position to another. We say that the continuous symmetry of the Lagrangian is spontaneously broken for this state.

It should be noted that there is no energy cost when the ball in Fig. 9.1 moves in the flat part of the potential in the brim of the Mexican hat, or when the s-electron of a free atom goes from one degenerate ground state to another. This is a general result, and is known as *Goldstone's theorem*. The Goldstone theorem in the nonrelativistic limit states that:

When a system has spontaneously broken symmetry, long-wavelength ($|\mathbf{k}| \to 0$) excitations exist, with a mode of frequency $\omega(\mathbf{k})$ which tends to zero in the limit $k \to 0$ (here k is the magnitude of the wave vector).

This zero-frequency 'Nambu-Goldstone' mode constitutes the limit of a continuous spectrum $\omega(\mathbf{k})$ of modes, the energy of which increases with increasing k.

Broken symmetry and gauge symmetry in particle physics and cosmology

Thou hast no faults, or I no faults can spy;
Thou art all beauty, or all blindness I.

Christopher Codrington

Broken-symmetry considerations are of central importance in particle physics, where we are particularly interested in the symmetries of the laws of Nature. The electroweak interaction is an example. This is the interaction which mediates the decay of a neutron into a proton, electron, and anti-neutrino. $SU(2)$ symmetry is involved here. When it is broken, the electron and the electron-neutrino become different. The $SU(2)$ symmetry of the electroweak interaction is firmly embedded in the laws of physics, but we do not observe it because the particular *state* in which we find our universe at present is not $SU(2)$-symmetry invariant.

Another example is that of $SU(3)$ symmetry in quantum chromodynamics (QCD). It is because of this broken symmetry that the red quark and the green quark and the blue quark are different. The present quantum-mechanical properties of the vacuum of our universe are such that, because of the broken $SU(3)$ symmetry, it distinguishes between the different states of quarks.

As discussed in the previous chapter on gauge symmetry, if the vacuum is not invariant under some symmetry, there must be a *field* that is making it so. That is, the field has a nonzero value even in its lowest-energy state. This is why the field breaks the symmetry: Being nonzero in its ground state, it affects different particles in different ways.

Such a field corresponds to the Mexican-hat potential shown in Fig. 9.1. At the peak of the potential the full cylindrical symmetry is manifest, but that is not the ground state. The ground state is at the brim of the hat, and a particle at any point in the brim does not experience the cylindrical symmetry.

In quantum field theory, fields undergo quantum fluctuations, and these fluctuations are associated with the spontaneous creation and annihilation of *particles*. And the curvature in the potential determines the *masses* of the particles. In Fig. 9.1, the curvature is zero when a particle moves *along* the brim of the hat, and it is nonzero when, starting from the brim, a particle moves radially towards or away from the centre. The zero-curvature-potential scenario involves *massless* particles. The Nambu-Goldstone theorem says that there will *always* be a massless particle when there is spontaneous breaking of a continuous symmetry. Such a particle is called a *Nambu-Goldstone boson*.

The massless particle corresponds to the zero-curvature oscillations of the field along the brim of the hat in Fig. 9.1. There is no restoring force in these oscillations. But there *is* a restoring force for oscillations in the radial direction, meaning that a *massive* particle must also exist. It is the *Higgs boson*.

Krauss (2017) has given an extensive and simplified description of the role played by gauge symmetry and broken symmetry in particle physics and cosmology. The various fundamental interactions behave differently because of some accident of our circumstances that hides their identical nature in the beginning of the universe. He gives the analogy of ice on a window on a cold day. The ice crystals point in different directions randomly. Suppose you were a civilization that lived on an ice crystal pointing in a particular direction. Then that would be a very special direction for you. It would be a fundamental aspect of your universe. The fact would be hidden from you that the direction of the ice crystal is irrelevant, and that the laws of physics are independent of that direction. What has happened to us is that we are living at present in a universe rather like that ice crystal. Due to some accident of our circumstances (a symmetry-breaking phase transition), some weird field filling all space froze into existence. The particles interacting with that field (Higgs field) behave like they have mass, whereas 'in reality' they would be massless if the field were not there. The photon does not interact with this field, so it has zero mass. This is how we understand why the weak nuclear interaction is different from the electromagnetic interaction (mediated by photons).

This explanation by Steven Weinberg was extended by him to the masses of all fundamental particles. The stronger a particle interacts with the Higgs field, the larger would be its mass observed by us.

Lastly a word about the gravitational interaction. It has not been cast in terms of a quantum theory yet. There is a minimum-energy state called *vacuum*, which undergoes quantum fluctuations allowed by the uncertainty principle. As a result, virtual particles and fields get formed and annihilated spontaneously. The virtual pairs of particles have energies, and they add up to an infinite value. Thus there is a problem of infinities. A possible way out for it has been found by postulating the presence of *supersymmetry*. An important feature of supersymmetry is that matter particles (fermions) and force particles (bosons) have a supersymmetric relationship: They are two aspects of the same thing. This means that each matter particle, say a quark, must have a force counterpart, and each force particle, say a photon, must have a matter counterpart. This can solve the infinities problem because it turns out that the infinities associated with force particles have a positive sign and those for matter particles have a negative sign, so they can cancel each other out to a large extent. But the calculations are so complicated that it has not yet been possible to substantiate such a theory of *supergravity*.

Another problem is that it has not been possible to find the partners expected to pair with the known fundamental particles. The supersymmetry theory predicts that the partner particles should be ~1000 times as massive as the proton, perhaps even more. Such massive particles have not been observed yet.

Condensed-matter physics

In quantum physics, two situations must be distinguished: Those in which the number of degrees of the system is finite, and those in which it is infinite. For the former case (in atomic physics, for example), there is always a symmetric ground state. This is because the ground state here is a superposition of *all* the allowed ground states.

When the number of degrees of freedom is infinite, global symmetry may be realized in two different ways. The first is the same as that for the finite-number-of-degrees-of-freedom case: The laws of physics are invariant under the symmetry operations, and the ground state is unique and symmetric. The second situation is that in which the ground state is asymmetric. It is this second (broken symmetry) situation that is responsible for the existence of crystals, ferromagnetism, superfluidity etc.

In condensed-matter physics, because of the possibility of phase transitions, the symmetry of any particular ground state of the system can be lower than the prototype symmetry G_0. For example, the ground state of a crystal of ice is less symmetric than that of liquid water from which it formed on freezing. The (partial) loss of translational symmetry on freezing causes *rigidity*, which means that the system in the crystalline phase fights displacements. But a uniform displacement of all the atoms costs no energy. And a slightly nonuniform displacement (signifying a low-frequency wave) costs very little energy. These low-frequency excitations [$\omega(k \to 0) \to 0$] are the acoustic phonons, or sound waves.

As postulated and formalized by Landau, the symmetry-lowering phase transition is heralded by the emergence of an order parameter, which breaks the symmetry of the parent phase. Goldstone (1961) argued that, subject to some qualifications and exceptions (Sethna and Huang 1992), one can generalize and say that whenever a continuous symmetry is broken, long-wavelength modulations in the symmetry direction should have low frequencies. These low-frequency modulations have *a symmetry-restoring effect* (in agreement with the symmetry-conservation law stated above in this chapter). They take the system from one broken-symmetry state to another by precisely those mappings which were symmetry operations before the symmetry-breaking transition occurred.

The fact that the ground state has broken symmetry means that the system is 'rigid': It costs energy to modulate the order parameter. Several examples can be given:

(i)   Crystals (products of broken (continuous) translational symmetry of the fluid phase) are resistant to shear deformations and low-frequency phonons.

(ii) In ferromagnetic crystals there is a breaking of the continuous rotational symmetry for the magnetisation. This leads to magnetic stiffness and *spin waves*.

(iii) Nematic liquid crystals pour easily, but resist bending. This orientational stiffness is because of the breaking of the full rotational symmetry. There are rotational waves in such liquid crystals for the same reason.

(iv) In superfluids, the broken gauge symmetry is responsible for the superfluidity, just as the broken translational symmetry in crystals is responsible for the shear moduli. The low-frequency Goldstone modes in superfluids are 'heat waves' (also called *second sound*): There is a periodic modulation of the temperature, rather like sound waves in a crystal.

But there are important exceptions to this general trend (Sethna 1992). Breaking of gauge symmetry is also involved across any superconductivity transition, but there are no corresponding low-energy excitations or Goldstone modes. This has to do with the Meissner effect, and is the equivalent of the Higgs mechanism propounded by particle physicists (Sethna and Huang 1992).

Another exception occurs in ordinary crystals. Two kinds of continuous symmetry are broken in going from a fluid to a crystal. One is the full translational symmetry, and this results in low-frequency Goldstone excitations, namely acoustic waves. The other is rotational symmetry. A crystal has only discrete rotational symmetry, compared to the continuous rotational symmetry of the fluid phase. *Where are the corresponding Goldstone modes?* Nobody knows for sure. Sethna and Huang (1992) draw an analogy between this situation and the Meissner effect in superconductors. In the latter, magnetic field is expelled from inside a sample on transition to the superconducting phase. Similarly, the registry among the various grains of any polycrystal is found to be such that, instead of distributing the strain caused by juxtaposing any two grains *throughout* the bulk of the grains, it is confined only to the grain-boundary region. It is as if the strain gets expelled from the bulk, and is confined only to the rather thin grain boundaries.

9.8 Discrete broken symmetries

Symmetry is also broken when a crystal passes from one solid phase to another. But now the initial symmetry is not a continuous symmetry: Crystallographic space groups are not continuous groups. Therefore, when there is a phase transition from one crystalline symmetry to another, the symmetry broken is a discrete symmetry. In such a situation there are only a finite number of degenerate ground states in the lower-symmetry phase, and there are, in general, no low-frequency excitations taking the system from one such state to another (because of the gaps separating the degenerate states).

Domains and domain walls are a consequence of broken symmetry. Domain walls can be regarded as the elementary excitations separating the distinct ground states, i.e. domains (Blinc and Zeks 1974; Chaikin and Lubensky 1995; Wadhawan 2000).

9.9 Broken symmetry and biology

> *Morphological asymmetry is one of those exceedingly rare characteristics of animals (and protists and plants) that has evolved independently many times.*
>
> Richard Palmer (2009)

> *The fact that the two sides of many animals' brains are not mirror images - particularly in humans - may ultimately*

> *help to explain the differences in behaviour between species and even among individuals.*
>
> Richard Saltus (2007)

According to the Wikipedia, 'Asymmetry is the absence of, or a violation of, a symmetry'. I think it would be more appropriate to use the word 'asymmetry' only for 'absence of symmetry', and use the phrase 'broken symmetry' for 'a violation of symmetry'. Admittedly, the two terms do have an overlap of meaning sometimes.

Both asymmetry and broken symmetry occur in biological systems. In Section 5.4 I discussed the evolutionary advantages of symmetry in biological processes. In Section 7.3 we saw how there is an interplay between symmetrization and dissymmetrization in the inanimate world. This interplay occurs in the animate world also, and has strong evolutionary underpinnings. The human body, for example, has a high degree of bilateral symmetry, but not *perfect* bilateral symmetry. For example, the heart is on the left side, and there is no mirror image of it on the right. Evolution of two hearts is not exactly a smart idea (even from the so-called emotional angle)! There may be serious coordination problems in the logistics of pumping blood, and a host of other problems as well (Palmer 2004). Similarly, the two sides of our brains are not mirror images (Saltus 2007).

A quest for the reasons for asymmetry and broken symmetry in biology takes us all the way down to particle physics. We have P-violation, CP-violation, but CPT symmetry. Here P stands for parity symmetry (or left-right mirror symmetry), C for charge-conjugation symmetry, and T for time-inversion symmetry. Parity is conserved for the electromagnetic, strong, and gravitational interactions, but is violated for the weak interaction.

Even the combined parity and charge-conjugation symmetry is violated: We observe a preponderance of baryons over antibaryons in our universe, and CP violation is one of the necessary conditions for this.

T-asymmetry in the macroscopic world is a given. Time 'flows' in one direction (the future), so we notice T-asymmetry everywhere at the *macroscopic* level. But at the microscopic level, time-symmetry in the equations of motion of fundamental particles in conceivable.

Combined CP and T symmetry, namely CPT symmetry, has so far not been observed to be violated in our universe.

In organic chemistry and biochemistry, the chiral nature (or left-right asymmetry) of a large variety of organic molecules is well known. Naturally occurring amino acids, which are the building blocks of our proteins, are almost all left-handed. Our bodies can digest only right-handed sugar molecules, and not left-handed ones. Naturally occurring DNA is predominantly right-handed; it is called B-DNA. The evolution of such asymmetry may have been either because of 'accidents of Nature' (things could have gone either way), or because of some deeper causes.

The presence or absence of symmetry in biological systems can be because of 'external' causes or 'internal' causes. External causes are associated with the environment. Random causes at an early stage of the development of an organism can also lead to a frozen asymmetry. For example, most people are right-handed and this results in a stronger right arm, etc.

Internal causes involve, for example, genetic factors.

The role of external and internal causes in the breaking of symmetry and the first appearance of asymmetry during evolution is a subject of substantial current interest, and its study throws light on both Darwinian and Lamarckian evolution (Palmer 2009).

Emergence and development of asymmetry is not something easily quantified and investigated. A good simplifying feature of some of the current comparative studies on morphological asymmetry is to specify the *direction* in which to specify the asymmetry. In terms of direction alone, three predominant types of asymmetry occur in species: *dextral*; *sinistral*; and *random*: Dextral asymmetry means all individuals are right-handedly or dextrally coiled, and sinistral means left-handed. Random asymmetry means an equal number of dextral and sinistral individuals.

A simpler subdivision is into *fixed* asymmetry and random asymmetry. The former means that all individuals are asymmetric in the same direction (right or left). The asymmetry may be of the external form, or of internal organs.

The direction of asymmetry is found to be fixed in some species, and random in others. Further, it is found to be inherited in some species, and acquired in others (Palmer 2009). Symmetry breaks when a system is unstable to fluctuations. This is true for both inanimate and animate systems.

Investigations on asymmetry in biology can pay rich dividends in terms of the insights obtainable into the very nature of Darwinian evolution (Palmer

2004, 2009). I quote Palmer (2009):

'Did animals with fixed morphological asymmetries evolve directly from symmetrical ancestors or from ancestors that exhibited random asymmetry? But this question actually represents a fundamental one in evolutionary biology: which comes first evolutionarily, mutations that yield novel phenotypes, or novel phenotypes, followed later by mutations that facilitate their development? In other words, from the perspective of left-right asymmetry, are mutations for rightness or leftness what generates new right- and left-sided phenotypes, or do new right- and left-sided phenotypes arise first, followed by mutations that stabilize development of rightness and leftness?'

9.10 The principle of local activity

As emphasized in this book, gauge symmetries are the stuff the laws of physics are made of (Zee and Penrose 2016). Gauge symmetries are *local* symmetries, rather than global symmetries. Localness is not something confined to just fundamental symmetries, but is very widespread in just about all natural phenomena. A *'local activity principle'* has been recognized which governs symmetry-breaking in particular, and the evolution of complexity in Nature in general (Mainzer and Chua 2013; Mainzer 2014). First enunciated by Chua (2005), the principle says that *no complex phenomena can emerge in any homogeneous media without local activity*. First enunciated in the theory of nonlinear electronic circuits in a mathematically rigorous way, it has been generalized and proven at least for the class of nonlinear reaction-diffusion systems in physics, chemistry, biology, and brain research. In general, the emergence of complex patterns and structures is explained by symmetry breaking in homogeneous media, which is caused by local activity.

Mainzer (2014) has given a mathematical definition of local activity. Here is an excerpt from this paper:

'The principle of local activity had originated from electronic circuits, but can easily be translated into other non-electrical homogeneous media. The transistor is an example of a locally-active device, whereby a "small" (low-power) input signal can be converted into a "large" (high power) output signal at the expense of an energy supply (namely a battery). No radios, televisions, and computers can function without using locally-active devices such as transistors. For the formation of complex biological and chemical patterns, Schrödinger and Prigogine demanded nonlinear dynamics and an energy source as necessary conditions. But, for the exhibition of patterns in an electronic circuit (i.e., nonuniform voltage distributions), the demand for nonlinearity and energy source is too crude. In fact, no patterns can emerge from circuits with cells made of only batteries and nonlinear circuit elements which are not locally active.

'In general, a spatially continuous or discrete medium made of identical cells interacting with all cells located within a neighbourhood is said to manifest complexity if the homogeneous medium can exhibit a non-homogeneous static or spatio-temporal pattern under homogeneous initial and boundary conditions. The principle of local activity can be formulated mathematically in an axiomatic way without mentioning any circuit models. Moreover, any proposed unified theory on complexity should not be based on observations from a particular collection of examples and explained in terms that make sense only for a particular discipline, say chemistry. Rather it must be couched in discipline-free concepts, which means mathematics, being the only universal scientific language.'

The principle of local activity is really fundamental in science, and can even be identified in quantum cosmology as symmetry-breaking of local gauge symmetries generating the complexity of matter and forces in our universe. The principle has applications in economic, financial, and social systems, as also in the emergence of new equilibrium states, symmetry-breaking at critical points of phase transitions, and risky acting at the 'edge of chaos'. (See Wadhawan (2017) for a basic introduction to complexity science, including a detailed discussion of 'edge of chaos').

10. Particle Physics, Cosmology, and the Search for New Symmetries

There's been a total transformation in the way we think about the fundamental physical world because of gauge symmetry, but no one knows about it. It happened between about 1960 and 1975, which from a theoretical perspective was the most revolutionary period of the 20th century in understanding the universe. All the laws of physics have this symmetry and it guides us in looking for new laws.

Lawrence Krauss

Broken-symmetry considerations occupy centre stage in modern theoretical physics (Trodden 2005; Schwichtenberg 2015; Zee and Penrose 2016). Particularly in particle physics, progress often means postulating and discovering a broken symmetry (usually a gauge symmetry). Why only a broken (or 'hidden') new symmetry? Because an unbroken symmetry would have been manifest anyway, and would thus be an *old* symmetry rather than a new symmetry. Broken symmetries in Nature are not always very obviously there; one has to go looking for them.

Our search for new broken symmetries is influenced by our knowledge that Nature is highly symmetric at very high energies. This symmetry is often broken or hidden at lower energies. This means that we can discover new laws of Nature by discovering new broken symmetries. Knowledge of such broken symmetries may ultimately enable us to unify all the four fundamental interactions, including the gravitational interaction.

10.1 The Standard Model of particle physics

But every known theory describing nature at a fundamental scale reflects some type of gauge symmetry. As a result, physicists now tend to think of symmetries of nature as fundamental, and the theories that then describe nature as being restricted in form to respect these symmetries, which in turn then reflect some key underlying mathematical features of the physical universe.

Lawrence Krauss (2017)

Four types of forces or interactions are known to be operative in our universe. The first is the *gravitational interaction*, or the gravitational force field. It is very weak, but is always present between any two particles or bodies. It is proportional to the product of the masses of the objects interacting. Since most of the celestial bodies are very massive, the gravitational force becomes very significant for them. Your weight is the gravitational force with which the rather massive Earth attracts you towards its centre.

Like charges repel, and unlike charges attract. Similarly, like magnetic poles (north-north or south-south) repel, and unlike magnetic poles (north-south) attract. Research showed that the electric interaction and the magnetic interaction are really two aspects of the same underlying phenomenon, so the term *electromagnetic interaction* was coined. This is the second of the four interactions.

The third is the *weak nuclear interaction*. It is operative, for example, inside the nuclei of radioactive materials, and is responsible for the emission of alpha-particles, beta-particles, etc. from inside such nuclei.

Lastly, we have the *strong nuclear interaction*, which is very strong but very short-ranged, and is responsible for the large binding energies of nuclei: A rather large amount of energy is required for extracting a proton or a neutron from inside the nucleus of an atom.

In quantum theory, not only are the elementary particles quanta of mass/energy, even the force fields or interactions are mediated by quanta. For example, when two electrons interact, the fundamental particle which mediates the interaction is the photon. One electron emits a ('virtual') photon and recoils in the process. The other electron absorbs the photon and also recoils. This back and forth exchange of photons constitutes the electromagnetic interaction.

The name 'Standard Model of particle physics' was given in the 1970s to a model for understanding the fundamental particles of our universe and the interactions among them. It not only included all that was known about fundamental particles at that time, but also made some correct predictions about new particles. A systematic classification of the particles is another of its accomplishments.

As mentioned above, there are two types of fundamental particles: The building blocks of matter, and the mediators of interactions among the particles. These two classes of particles differ in a quantum parameter called *spin*.

The spin of any fundamental particle can be either an integer (in units of $h/2\pi$, h being the Planck constant), or a half-integer. That is, it is either integral (0, 1, 2, 3, ...), or half-integral (1/2, 3/2, 5/2, ...). The former class of particles are called *bosons*, and the latter are called *fermions*. Fermions, by definition, obey Femi-Dirac statistics, and bosons obey Bose-Einstein statistics. *The particles constituting matter are all fermions, and the particles mediating interactions are all bosons.*

Whereas the Pauli exclusion principle of quantum mechanics prevents fermions from occupying the same quantum state, there is no such restriction for bosons. Thus, unlike fermions, bosons *can* have the same set of quantum numbers. [A quantum state is specified by a set of quantum numbers. Spin is an example of such a quantum number; there are many others.]

Electrons, protons and neutrons are fermions, with spin 1/2. By contrast, photons are bosons, with spin 1.

Gauge symmetry forms the basis of the Standard Model. The present version of the model is able to account for three of the four fundamental forces or interactions of Nature (electromagnetic, weak nuclear, and strong nuclear). It has not been possible so far to include the gravitational interaction in it.

The familiar electric charge (its conservation was discussed in Chapter 8) is relevant to the electromagnetic interaction. Other conserved 'charges' have been postulated for dealing with the weak nuclear interaction and the strong nuclear interaction ('weak isospin' charge, 'colour' charge, etc.). The Standard Model is a *model* and not a *theory*. In it one models the dynamics of the various conserved 'charges' by associating the respective gauge degrees of freedom with them, and invoking local symmetries for the transformations of those degrees of freedom which are dual to the 'charges'. The gauge degrees of freedom are not *observables*, but the model enables us to draw conclusions about how such 'charges' must interact and evolve.

All laws of Nature are quantum-mechanical laws. Classical laws are only approximations or limiting cases of the laws of quantum mechanics, adequate in many day-to-day situations. At very small sizes (e.g. those of the fundamental particles), quantum effects are important. Therefore, it is necessary to have a quantum-mechanical description of the particles that constitute matter, as well as of the four possible interactions among them.

We need a quantum field theory for all the elementary particles, both matter particles (fermions) and interaction particles (bosons).

The progress made in quantum field theories has brought forth again and again the importance of gauge transformations. Gauge symmetry puts restrictions on the laws of physics. The changes induced by a gauge transformation must cancel one another out when expressed in terms of observable quantities. In fact, it has turned out that in physics all the fundamental interactions or forces arise from the constraints coming from the always local gauge symmetries. The forces arise because the local gauge transformations vary from point to point in space, and from moment to moment in time. The forces are mediated by *gauge bosons*, the nature of which is determined by the gauge transformations. The work on gauge symmetry during the 20th century has resulted in the highly successful Standard Model.

Before Maxwell did the consolidation work regarding the electromagnetic interaction (through his celebrated equations, which also resulted in the identification of 'displacement current'), it was Faraday who first recognized the presence of a *force field* for any interaction or 'action at a distance' between two or more entities. It is now accepted that *all* forces are transmitted through fields. Electromagnetic fields propagate through space at the speed of light. In fact, light itself is an electromagnetic wave.

Maxwell's theory of the electromagnetic interaction was a *classical* theory. So also was Einstein's theory of the gravitational interaction, namely the general theory of relativity. Quantum effects become very significant at sub-atomic length scales. Moreover, the early universe (at and immediately after the Big Bang) was also of very small dimensions. We therefore need a quantum formulation for all the four interactions.

The first interaction to be quantized was the electromagnetic interaction, and the resulting field of research is called *quantum electrodynamics* (QED). Two particles interact because each creates a field around itself, which is felt by the other particle. As mentioned above, in the quantum version of the fundamental interactions even the fields are quantized and associated with corresponding elementary particles. The photon is an example of the quantization of the electromagnetic field.

Richard Feynman, who played a major role in the development of QED, had also formulated a *sum-over-histories* version of quantum mechanics. He used his famous *path-integrals* approach for working out the details of QED.

There is not just one way in which a photon may be emitted by one electron and absorbed by another. All possible modes or histories of emission and absorption must be considered and summed up vectorially. This is because in quantum mechanics a particle can be everywhere in space at the same time, so we cannot compute a unique quantum trajectory the way we compute a classical trajectory. All conceivable trajectories or histories must be summed up. But when this sum over all alternative histories was carried out, the problem of '*infinities*' was encountered: There are infinitely many histories to sum, so the QED theory ended up calculating an infinite mass and an infinite charge for the electron, which was an absurdity.

Feynman not only established the subject of QED, he also introduced the vitally important and much-used '*Feynman diagrams*'. Another fundamental development by him was the '*renormalization*' procedures he introduced to get over the problem of infinities mentioned above. In the early stages of the work, the renormalization involved subtracting infinite but negative terms from the sum over histories such that what were left were finite numbers. This was not a very elegant theory to begin with, because it allowed just about any value for the electron charge and mass; the theory did not *predict* their values. But the saving grace was that once you had set the adjustable parameters, namely the charge and the mass of the electron, equal to the experimentally known values, all further predictions of the theory were borne out by experiment to a very high degree of precision. An example is the correct prediction of what is called the *Lamb shift*, a small change in one of the energy states of the hydrogen atom.

The laws of Nature remain invariant under the gauge transformation involving a local reversal of the signs of all the charges. This requires the existence of an electromagnetic field that is governed precisely by the Maxwell equations. This gauge invariance completely determines the nature of electromagnetism and QED. *QED is the most precise and preeminent quantum theory of the 20ᵗʰ century.*

Encouraged by the success of the theory of QED, physicists attempted to formulate quantum field theories for the other three fundamental interactions also. The idea of gauge invariance was sought to be extended from QED to the other interactions also. Unique 'infinities problems' were encountered for each interaction. The photon is its own antiparticle, so the infinities problem had to be tackled in a special way. But for other interactions, the virtual particles involved in the summation over all histories have distinct antiparticles. So the possibility existed that the infinite and positive contribution to the summation coming from particles can get cancelled both in magnitude and sign from the contribution from antiparticles. ['Charge'-conservation symmetry demands that in any virtual emission process, the virtual particle and antiparticles must appear and disappear simultaneously.]

Alongside this activity, work was also going on for unifying all the interactions into a single interaction. It had been realized that as we go up the energy scale, the interactions would merge one by one, so that at high-enough energies only one interaction would prevail. The highest-energy scenario, of course, was what transpired at and soon after the moment of the Big Bang, when there was only one fundamental interaction in operation. As our cosmos expanded and cooled, *symmetry-breaking transitions* occurred, resulting in the successive emergence of the four interactions we know at present.

That unification is the correct approach became clear when attempts were made to formulate a quantum field theory of the weak nuclear interaction. It was found that the nuisance of infinities could not be gotten over by a renormalization procedure, except when the electromagnetic interaction and the weak nuclear interaction are treated as one single interaction, since called the *electroweak interaction*. Glashow, Salam and Weinberg were awarded the Nobel Prize for this work in 1973.

QED is determined completely by the gauge invariance under the operation that reverses the signs of all the charges of Nature. In analogy with this, what would happen if the identities of all the neutrons and protons were interchanged, in different manners at different points of space? Naturally, some new field will have to be defined at each point for accounting for and annulling the effects of this gauge transformation. Yang and Mills asked this question and wondered whether such a field would be a quantum field, with particles associated with it playing a role in determining the forces between neutron and protons. It was postulated that such a gauge symmetry would be associated with the conservation of the isotopic spin 'charge'.

Unlike in QED, switching the signs of isotopic-spin 'charges' of protons and neutrons in a nucleus replaces a (neutral) neutron with a (charged) proton, and a charged proton with a neutral neutron. Therefore, the new field ('*Yang-Mills field*') introduced for cancelling out the effects of the local gauge transformations, ensuring that the underlying physics remains unchanged, must itself be a charged field. But being charged, such a field would have to interact with *itself*, unlike the situation with photons, which are charge-neutral.

This brought in some complications. For starters, one would need *three* such fields, not one: One positively charged, one negatively charged, and one neutral. Therefore, whereas the electromagnetic field in QED is a *vector* field, this new field would be a *tensor* field, represented by a matrix. Matrices have the property that, in general, a product of two matrices depends on the order in which they are multiplied: AB ≠ BA. So the underlying gauge symmetry (*Yang-Mills gauge symmetry*) is a non-Abelian symmetry. Further, the electromagnetic interaction (as also the gravitational interaction) is a long-range interaction. But the weak nuclear interaction (as also the strong nuclear interaction) is a short-range interaction. It operates only over intra-nuclear distances. Long-range interactions are mediated by massless quanta; photons in the case of electromagnetism, and gravitons in the case of gravity. The nonzero rest-mass nature of the particle(s) mediating the weak nuclear interaction was another complication. A credible quantum theory for a massive mediating particle had not been formulated till then.

The next major advance came from Julian Schwinger. In 1957 he made the suggestion that the gauge symmetry of QED may be just a part of a larger gauge symmetry in which new gauge particles (in addition to the photon of QED) mediate the weak nuclear interaction involved in the decay of a neutron to a proton. This was the genesis of the unification of the electromagnetic interaction and the weak nuclear interaction into what we now call the electroweak interaction. The Yang-Mills theory was back in the reckoning.

Schwinger tried to account for the fact that the weak nuclear interaction is far weaker than the electromagnetic interaction by stipulating that the new gauge particles are probably *very* heavy, ~100 times heavier than protons and neutrons, so that the range of the interaction they mediate would be smaller than even the size of a proton or a neutron. The intrinsic strength of coupling of such short-range fields with electrons and protons on small scales would be comparable to electromagnetic fields over such length scales, and yet be much weaker on the scale of nuclei.

But there were problems with this model. Why should the charged gauge particles have a huge mass while the photon was massless? Nevertheless, Schwinger's student Sheldon Glashow continued to pursue the idea further. He took note of the additional fact that the weak nuclear interaction seemed to apply only to *left-handed* electrons and neutrinos, whereas the electromagnetic interaction was independent of the handedness. Some conceptual breakthrough was needed.

Renormalization of the quantum field theory for the strong nuclear force could be carried out on its own, and the theory is known as *quantum chromodynamics* (QCD). According to it, protons, neutrons and some other fundamental particles are composed of still more fundamental particles called *quarks*. Murray Gell-Mann was awarded the Nobel Prize for his work on quarks.

Quarks come in three *colours* (nothing to do with the usual meaning of colour): red, green, and blue, along with the respective *anticolours*. The quarks cannot exist as free, stable particles. Only those combinations of them can exist as free particles which do not have a net colour. For example, a colour and its anticolour cancel, giving a neutral net colour. Composite particles in which this occurs are what are called *mesons*. Another possibility is that all three colours (one each), or all three anticolours (one each), occur together in a composite particle. The name for such a composite particle is *baryon*. Protons and neutrons are examples of baryons.

In addition to colour, quarks have quantum parameters like 'up' (u), 'down' (d); 'charm', 'strangeness'; and 'top', 'bottom' (do not pay attention to the literal meanings of such words; they are just labels). Two up quarks and a down quark make a proton, and two down quarks and one up quark make a neutron.

QCD has the *asymptotic freedom* feature: The quarks interact via the strong force, which behaves like a piece of stretched rubber: The force is small at shorter distances, and becomes stronger (like a taut rubber) when the distance between the interacting quarks increases.

Various attempts have been made for uniting the electroweak interaction with the strong nuclear interaction. These 'grand unification theories' (GUTs) have not been particularly successful. Most of them predicted that the proton is not an eternally stable particle, and should decay with a 'half-life' of $\sim 10^{32}$ years. Experimental verification of proton decay has not been confirmed unambiguously.

So, we have the Standard Model, in which there is the unified electroweak interaction, a separate strong nuclear interaction, and the (most problematic) separate gravitational interaction. 'Most problematic' because there is still no quantum version of the gravitational interaction, or a theory of *quantum gravity*. One reason for this is the Heisenberg uncertainty principle of quantum mechanics. The principle applies to various pairs of 'conjugate parameters'; e.g. position and momentum; or energy and time. Another such pair is the value of a field and its rate of change. The gravitational field is extremely weak by nature, meaning that a very large uncertainty in its rate of change is readily permitted by the uncertainty principle. This means that *there is no such thing as empty space*: Since a large uncertainty in the rate of change is permitted by the uncertainty principle (for regions where the gravitational force is very weak), an empty space would not be able to remain empty for a significant duration. There is a minimum-energy state called the *vacuum*, and the vacuum is subject to *quantum fluctuations* permitted by the uncertainty principle: 'Virtual' pairs of particles, and of fields, get formed and annihilated spontaneously all the time. The virtual pairs have energies, and they add up to an infinite value. A problem of infinities again. An apparently insurmountable problem in this case, though, is that it is not possible to apply the renormalization trick, because there are not enough adjustable parameters here (unlike charge and mass in the case of the electromagnetic interaction).

Gauge invariance necessitates the existence of particles called *gauge bosons* for the corresponding quantum fields. Gauge bosons are massless particles that are associated with the vector potential. The Standard Model has $SU(3) \times SU(2) \times U(1)$ gauge symmetry, which demands the existence of eight gluons for mediating the strong interaction, three gauge bosons (the two $W^{\pm}$ bosons, and the Z boson) for mediating the weak interaction, and the photon for mediating the electromagnetic interaction. These gauge bosons all have spin, as do matter particles.

There has been an unexpected fallout of the GUTs mentioned above. Ideas from the GUTs have found applications in cosmology. A certain *scalar field* (the *Higgs field*) in the GUTs has provided a strong basis for supporting the idea of *cosmic inflation* postulated for the neonatal universe: The gravitational interaction arose 10^{-43} seconds (the *Planck time*) after the Big Bang. The other three interactions were still in a 'grand-unified' state. The spontaneous emergence (on further cooling of the universe) of a GUT scalar field 10^{-35} s after the Big Bang broke the *GUT symmetry*, and new interactions appeared one by one, the first to appear being the strong interaction.

Prior to the emergence of new interactions, the GUT scalar field had also caused a major and very rapid *inflation*, doubling the size of the baby universe every 10^{-34} s. The inflation was so rapid that the quantum fluctuation just 10^{-20} the size of a proton could grow to a size of 10 cm in just 15×10^{-33} s. This is how the GUT scalar field kick-started the growth of the universe, *preventing a gravitational-pull-mediated collapse of the quantum fluctuation that had initiated the universe.*

In the Standard Model there are 12 elementary particles of spin 1/2: six quarks and six *leptons*. Further, each such fermion has a corresponding *antiparticle*. For example, the antiparticle of an electron is a particle called the *positron*. Thus, there are 24 fermions in all.

The six quarks are: up, down, charm, strange, top, and bottom quark. And the six leptons are: electron, electron-neutrino, muon, muon-neutrino, tau, and tau-neutrino. The quarks interact via the strong nuclear interaction. They form 'colour-neutral' composite particles called *hadrons*. The hadrons are either baryons, or mesons. The baryons (protons and neutrons) contain three quarks, and the mesons comprise of a quark and an antiquark.

It was believed earlier that there are three types of gauge bosons: (i) photons; (ii) W^+, W-, and Z particles; and (iii) gluons. Out of these, photons and gluons are massless (zero rest-mass) particles. Photons mediate the electromagnetic interaction. There are eight types of gluons, and they mediate the interaction among quarks. The W^+, W-, and Z gauge bosons have nonzero rest mass, and they mediate the (short-ranged) weak-nuclear interaction among quarks and leptons. But now a fourth type of gauge boson must be added, namely the *graviton*. It is expected to be massless, and have spin = 2.

The Higgs field permeates all space. The *Higgs particle* (or *Higgs boson*) is the quantum of this field, and it has nonzero rest mass. The Higgs field is a scalar field (i.e., it has zero spin). The Higgs field interacts, with different strengths, with the various particles. Particles that interact more strongly with it experience more resistance or drag to their motion, and thus are more massive. Some particles, such as photons, do not interact with the Higgs field at all, and are therefore massless (zero rest mass). In this way, the mass of everything is determined by the existence of the Higgs field.

Fig. 10.1 is a pictorial depiction of Standard Model.

The Higgs field is a part of the theory of the electroweak interaction (see Krauss 2017). It has a role in breaking the symmetry between the electromagnetic interaction and the weak interaction. This symmetry existed at very high temperatures, and was broken spontaneously as the universe cooled a little and the Higgs field appeared. Particles with nonzero rest mass arose when this symmetry was broken by the emergence of the Higgs field.

The physical laws governing the Higgs field have high symmetry. Yet the present actual state of the Higgs field has a spontaneously-broken or lower symmetry. The field takes a particular value at low energies, and pervades all space, rather like the energy associated with vacuum in quantum field theory. Particles acquire their masses by interacting with the pervading Higgs field.

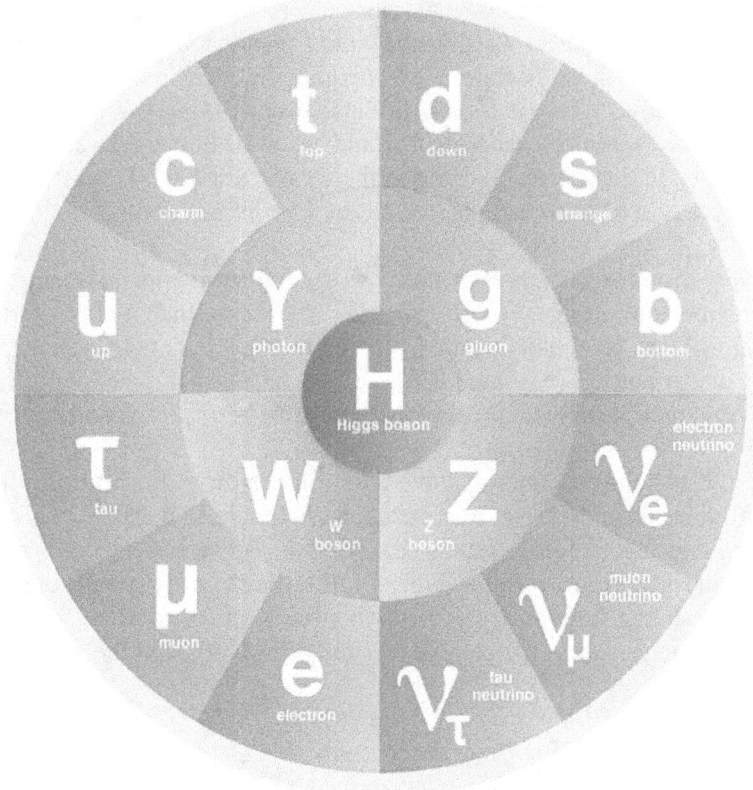

Fig. 10.1 Depiction of the Standard Model as a kind of periodic table of the elements for particle physics. But instead of listing the chemical elements, it lists the fundamental particles that make up the atoms that make up the chemical elements, along with any other particles that cannot be broken down into any smaller pieces. (http://www.symmetrymagazine.org/standard-model)

10.2 Beyond the Standard Model

> *. . . the discovery and application of this mathematical symmetry, gauge symmetry, has allowed us to discover more about the nature of reality at its smallest scales than*

any other idea in science.

Lawrence Krauss (2017)

As discussed above, the uncertainty principle is one reason why it has not been possible so far to formulate a satisfactory quantum theory of gravity. When we speak of vacuum in quantum physics, we really mean a space which has a certain minimum-energy state. This state is subject to '*quantum fluctuations*', which means that pairs of (virtual) particles can make momentary appearances (within the limits prescribed by the uncertainty principle), and then disappear by merging into each other. There are infinitely many such virtual pairs possible, each having energy. But if the vacuum state has infinite energy, it would curve the universe to an infinitely small size, according to the general theory of relativity. This is not what actually happens, so we are plagued by another infinity problem.

In 1976 the idea of *supersymmetry* was put forward in this context. In supersymmetry theory, force particles (bosons) and matter particles (fermions) are symmetry-related, or rather *super*symmetry-related: For every matter particle (e.g. a quark) there must be a 'super' force particle, and for every force particle (e.g. a photon) there must be a 'super' matter particle.

This idea has the potential to solve the infinities problem. It turns out that the infinities from matter-related virtual particles are all negative, while they are positive for all force-related virtual particles, so they can cancel each other out. The necessary calculations are difficult to carry out, but many believe that the notion of *supergravity* which emerges when we invoke supersymmetry has the potential to unify gravity with the other three interactions.

At and soon after the Big Bang the energies were so high that there was only one fundamental interaction. As our universe cooled and expanded, symmetry-breaking transitions occurred and different interactions arose one by one. Concomitantly, new fields, as also matter, arose as a result of these transitions. Since there was only radiation, and no matter, to start with, the present observed distinction between matter particles (fermions) and field particles (bosons) must also be a result of a broken symmetry. We call it *supersymmetry*. If supersymmetry can be restored (at high enough energies), the distinction between fermions and bosons would vanish at those energies.

Supersymmetry is described in a certain higher-dimensional superspace, four of these dimensions being for the spacetime coordinates we perceive in our world. The additional dimensions and coordinates are denoted by

θ_i, $i = 1,2,....$ The extra coordinates have the property that they are anticommuting numbers; i.e., $\theta_1\theta_2 = -\theta_2\theta_1$. Supersymmetry operations are continuous coordinate transformations in this supersymmetric hyperspace.

The most important new symmetry emerging from the supersymmetry description of Nature is that for every particle with spin J (a boson), there must be another particle with spin $J \pm 1/2$ (a fermion). There would be a degeneracy if this supersymmetry were not broken, and the masses of the two partner particles would be equal. But since it is broken, the masses are different. Detailed experimental verification of such conclusions is awaited.

The important point is that recognition or postulation of a new broken symmetry enables us to make testable scientific predictions.

The idea of supersymmetry had actually originated earlier when *string theory* was being formulated. The Standard Model does not include quantization of the gravitational interaction, and its unification with the other three interactions. This means that additional broken symmetries need to be postulated and verified, with an attendant increase in the number of dimensions of the hyperspace in which the symmetry transformations operate. String theory attempts to do that. It involves an extension of the conceptual framework of local quantum field theory.

In string theory, the elementary particles are envisaged, not as points, but as patterns of vibration that have length but no width or height ('strings'). There are several string theories, and they are consistent only if spacetime has 10 dimensions, rather than 4. We see only four dimensions because the other six are curved up (or curled up) into a space of very small size.

An analogy will help understand this. Consider a straw you use for drinking lemonade. Its surface is 2-dimensional: We need two numbers or coordinates for specifying the location of any point on it. But if the straw is extremely thin (say a million-million-million-million-millionth of a centimetre), it is practically 1-dimensional; the other dimension has just curled up into near-nothingness in terms of visibility. The extra dimensions in string theory are said to have curled up into '*internal space*'.

An awkward problem faced in early days was that there appeared to be at least five different string theories, and millions of ways in which the extra dimensions could be curled up. Then, in the early 1990s, '*dualities*' were discovered: It was realized that the different string theories, as also the myriad ways of curling up the extra dimensions, are simply different ways of describing the same phenomena in four dimensions. It was also found that supergravity is also related to the other theories in this manner.

Many experts are now convinced that the five string theories, as also supergravity, are merely different approximations to a more fundamental theory called the *M-theory*, each valid in different situations.

M-theory involves 11 dimensions instead of 10. It is this extra dimension which unifies the five string theories. Moreover, M-theory allows for not just 1-dimensional strings, but also point particles, 2-dimensional membranes, etc., all the way up to 9-dimensional entities (*p-branes*, with *p* running from 0 to 9). M-theory is the unique supersymmetric theory in 11 dimensions.

A crucial feature of M-theory is that its mathematics restricts the ways in which the dimensions of the internal space can be curled. Thus the theory comes up with *unique* (rather than arbitrary) values for the fundamental constants and the 'apparent' laws of physics corresponding to any particular mode of curling (see below).

10.3 Origin of our universe

But how did our universe arise anyway? Our universe is known to be expanding. This means that if we extrapolate backwards in time, there must have been a moment when our universe was extremely small, almost a point. It has been estimated that that was the case ~13.7 billion years ago. There was a Big Bang, and our universe has been expanding ever since then.

The Big Bang point is taken as the zero of spacetime. It is a 'singularity' because certain quantities become infinite at this point in Einstein's equations of general relativity. Therefore Einstein's theory is applicable only a little after the Big Bang, and not at the singularity.

Evidence in support of the Big Bang model comes from many sources. One is the observation of the cosmic microwave background radiation (CMBR). The observed distribution of this radiation is quite uniform, but not *very* uniform. In fact, the minute structure it has is responsible for the evolution of galaxies etc. (Hawking and Mlodinow 2010).

How could our universe get created spontaneously out of nothing? Was there a violation of the principle of conservation of energy? No. We can explain the emergence of positive energy (radiation or/and matter) out of nothing if there is a *simultaneous* emergence of a balancing amount of negative energy. This negative energy arose because the Big Bang was accompanied by the emergence of the gravitational interaction, which is an attractive interaction. Any attractive interaction engenders a negative contribution to the total energy because it takes positive energy to break free from the binding force of the attractive interaction.

There is no reason why only one universe, namely ours, should emerge out of nothing. The M-theory tells us that a very large number of universes can emerge, and go their separate ways.

Cosmological observations make it necessary for us to postulate a very brief period of very rapid '*inflation*' soon after time-zero, much more rapid than even the speed of light (this is possible because the expansion of space itself *can* be faster than the speed of light). It is this inflation which was the 'bang' in the Big Bang.

But there is a problem. For explaining inflation and its aftermath, some very special conditions must exist at time-zero. The model proposed in Hawking and Mlodinow (2010) for the creation of the universe is such that this problem gets eliminated. How?

Time-zero was the moment when spacetime came into existence. We know from Einstein's general theory of relativity that the gravitational interaction can be viewed as a warping of spacetime, so when gravitation came into existence, so did spacetime. And one reason the time dimension gets mixed with the space dimensions is that matter and energy warp time. This mixing is a key element for understanding the beginning of time.

Some earlier research work by Hawking and Hertog (2002, 2006) had established that when we add the effects of quantum theory to the general theory of relativity, in extreme cases warpage can occur to such an extent that time behaves like another dimension of space. Thus in the early universe, when it was small enough to be governed by both general relativity and strong quantum effects, there were effectively four dimensions of space and none for time. Time as we know it did not exist when we extrapolate backwards in time towards the very early universe. So how did time begin? I quote from Hawking and Mlodinow (2010):

'Suppose the beginning of the universe was like the South Pole of the Earth, with degrees of latitude playing the role of time. As one moves north [from the South Pole], the circles of constant latitude, representing the size of the universe, would expand. The universe would start at a point at the South Pole, but the South Pole is much like any other point. To ask what happened before the beginning of the universe would become a meaningless question, because there is nothing south of the South Pole. In this picture spacetime has no boundary - the same laws of nature hold at the South Pole as in other places. In an analogous manner, when one combines the general theory of relativity with quantum theory, the question of what happened before the beginning of the universe is rendered meaningless.'

The term '*no-boundary condition*' is used for the idea that the histories of the universe are closed surfaces without a boundary (in an appropriate hyperspace).

Since the origin of the universe was a quantum event, Feynman's sum-over-histories formulation for going from spacetime point A to spacetime point B occupies centre stage. But we have knowledge only about the present state of the universe (point B), and we know nothing about the initial state A. Therefore, according to Hawking, we can only adopt a '*top down*' approach to cosmology, wherein every alternative history of the universe exists simultaneously, and the histories relevant to us are only those which satisfy the no-boundary condition and which, when summed up, give us our present universe (point B).

The picture that emerges is that the universe, or rather a whole lot of them, appeared spontaneously (and the M-theory allows for $\sim 10^{500}$ of them). Most of these multiple universes were not relevant to us because their 'apparent' laws were not conducive to our emergence and survival.

What enters the sum over histories relevant to us in not just one universe. Although one particular universe with a completely uniform and regular history does have the highest relative probability amplitude and therefore contributes the maximum to the Feynman sum, several others, which have slightly irregular or deviant histories but still significant probability amplitudes, also contribute to the sum over histories. This should account for the slight nonuniformities during the inflation era, as evidenced by the CMBR plot. These irregularities were important for the emergence of galaxies. 'We are the product of quantum fluctuations in the very early universe.'

Is this theory testable? Yes. The no-boundary condition implies that the probability amplitude is the highest for histories in which the universe starts out completely smooth. And it is somewhat smaller for universes which are slightly irregular by comparison. Starting from the M-theory one can work out the details of how the CMBR pattern should look, and then compare with detailed and accurate experimental observations.

The M-theory offers $\sim 10^{500}$ possibilities of start-up universes. We have to single out those which correspond to the curling up of exactly those dimensions which we find to be the case for the universe we inhabit. The narrowed-down choice should also be such that it predicts conditions which make it possible for inflation to start and proceed exactly the way it actually did for our universe. Of course, we select those histories which reproduce not only the observed mass and charge of the electron, but also other such observed fundamental parameters.

In fact, one can work backwards the way it was done for tackling the problem of infinities in QED. There is a whole 'zoo' of elementary particles for which we have knowledge of mass, charge, spin, etc. from experiment. We can plug these numbers into the M-theory for cutting down the choice of adjustable parameters in the theory. It would be interesting to see if the M-theory does indeed get validated by experiment. If it does, that would be the most profound vindication of our perception that *symmetry is supreme.*

11. Latent Symmetry, Potential Symmetry, and the Symmetry Composition Principle

> *As far as I see, all* a priori *statements in physics have their origin in symmetry.*
>
> Weyl, *Symmetry*

11.1 Latent symmetry and potential symmetry

We have seen in earlier chapters that the symmetry of an entity is often a manifestation of *equality* or *equivalence* among different parts of the entity (Lederman and Hill 2004/2008). Because of this equivalence, when a symmetry transformation is applied the entity transforms back into itself, as if no transformation has been applied.

An object is said to possess *latent* symmetry (LS) if, when two or more copies of it are brought together or superimposed in a certain special way, the symmetry of the composite object is *higher* than what was expected from the symmetry of the original object and the symmetry ('placement symmetry') of the particular configuration chosen for constructing the composite object from the equal parts (copies) (Wadhawan 1987, 2000; Litvin and Wadhawan 2001, 2002; Litvin, Wadhawan and Hatch 2003). This definition will be made rigorous in Chapter 12 in the language of group theory.

The original object can be considered as the *building block* (BB), from which the composite object gets constructed by an assembly or juxtaposition or superposition of a number of identical copies of the BB. Fig. 11.1 illustrates this by taking an isosceles triangle as the BB.

Two such isosceles triangles are shown in Fig. 11.1(a), juxtaposed in a special way. The one on the left of the vertical line is the original, and the one on its right is a copy, obtained by reflection across the vertical line. The composite figure is a rhombus. It has two mirror planes, or rather mirror lines, of symmetry, one passing through the horizontal diagonal (let us denote it by m_y), and the other passing through the vertical diagonal (m_x). The symmetry element m_y is a property of the original isosceles triangle, as also of its copy. And m_x is a consequence of the fact that we have *chosen* to juxtapose the two triangles in a special way depicted in Fig. 11.1(a). So, there are no surprises so far. Everything is accounted for.

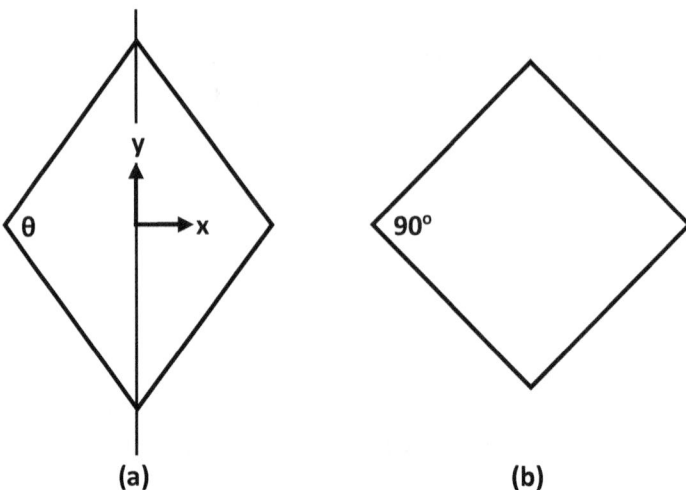

(a)   **(b)**

Fig. 11.1 (a) A composite object (rhombus) made up by joining two equal (mirror-reflected) isosceles triangles having an apex angle θ different from 90°. (b) Same as (a), except that the apex angle is now 90°. It has the symmetry of a square, which is higher than that of a rhombus.

The surprise comes when we take an isosceles triangle for which the apex angle θ has the special value 90°. We then get Fig. 11.1(b), which is a *square* rather than a rhombus. The square still has the m_y symmetry and the m_x symmetry, just as the rhombus on the left has. But it also has an additional symmetry axis usually denoted by the symbol 4; i.e., if we rotate the square by an angle 2π/4 (or 90°), we get back the same square. *This is a symmetry element not present in the rhombus.*

We say that an isosceles triangle having an apex angle 90° has a *latent* symmetry (namely a 4-fold axis of rotation symmetry) which becomes *manifest* symmetry when two such equal triangles are superimposed or juxtaposed in a special way shown in Fig. 11.1(b). The symmetry is latent, and becomes manifest only when the composite object is constructed in a special way, i.e., only when there is an appropriate placement symmetry.

There are other examples. A crystal is a composite object comprising of a large number of equal parts, namely the unit cells. If a crystal has no symmetry other than the mandatory translational symmetry, then the unit cell itself is the BB. This corresponds to a crystal with a space-group symmetry P1.

Next consider a crystal with a somewhat higher symmetry, say P2. The 2 in this symbol denotes the fact that the crystal has 2-fold rotational symmetry. For this case the BB can be identified as a portion of the unit cell which has only half the volume of the full unit cell. This is because the other half of the unit cell is equivalent (or symmetry-related) to the first half via the 2-fold rotation operation of symmetry. The asymmetric unit is the BB here, from which the full crystal can be generated by applying the symmetry operations of the space group of the crystal.

Let us assume here without loss of generality that the BB or the asymmetric unit consists of only one molecule, or a fraction of that molecule. If there are more than one molecules in the BB, then the whole collection of them can be treated as a single supermolecule.

What decides whether the crystal grows to have symmetry P2, and not P1? Obviously, it has to do with the shape of the molecule in the BB, and the nature of the force field around it which is responsible for the interactions among the various copies of the BB in the fluid from which the crystal has grown, unit by unit (see Wadhawan (2000) for a discussion of how the final macroscopic symmetry of a growing crystal emerges, starting from the symmetry of a cluster of two or more molecules).

If the symmetry of the bulk crystal is only P1 it means that, through a process of trial and error, the set of identical molecules (equal objects) constituting the crystal has found a least-free-energy configuration such that the assembly has translational symmetry, but no directional or point-group symmetry.

If the symmetry of the crystal is higher than P1, say P2, it means that the shape and the governing interactions among the identical molecules (the BBs) are such that the least-free-energy configuration of the resultant crystal is one in which there are *two* BBs in the unit cell, *and which are so oriented with respect to each other* that we can obtain the coordinates of the atoms in one BB by a 2-fold rotation of the other BB about a suitably located axis of rotation.

It is important to realize that this 2-fold symmetry of the crystal is entirely a consequence of tendencies and forces present in the BB or the molecule. The shape of the molecule is such, and the nature of the force fields emanating from it are such, that we can say that 2-fold symmetry was inherently present as a *potential* symmetry (*latent or/and placement symmetry*), and became manifest when two such identical molecules got a chance to interact with one another repeatedly till they could together settle down to a unique energy-minimizing mutual orientation and disposition with a manifest 2-fold symmetry.

Let us dwell some more on the two examples of the molecules considered above, one crystallizing with space-group symmetry P1, and the other with space-group symmetry P2. Taken in isolation, none of these two kinds of molecules has any symmetry at all. The translational symmetry in crystals of the first molecule is a *potential* symmetry of the molecule, which became manifest symmetry when its crystal got formed. This molecule has no other potential symmetry.

The second molecule, which spontaneously forms crystals of symmetry P2, has two potential symmetries. One became manifest as the translational symmetry of its crystal. The other as the 2-fold rotational symmetry of the crystal.

11.2 The distinction between potential symmetry and latent symmetry

Latent symmetry is potential symmetry as well, but potential symmetry may or may not be latent symmetry. Before we discuss this further, let us make a distinction between *natural symmetry* and man-made or *artificial symmetry*.

Natural symmetry is what molecules and crystals possess spontaneously. We cannot force a molecule to have symmetry of our choice, nor can we force a collection of molecules to form a crystal having our choice of symmetry. The term 'potential symmetry' is relevant only in the context of spontaneously occurring or *natural* processes.

But we humans can also create objects of symmetry. This is artificial symmetry or man-made symmetry. Fig. 11.1(a) provides an example. The isosceles triangle has the inherent symmetry m_y. And the rhombus created from two such triangles has the additional symmetry m_x. This latter symmetry has arisen because we chose to place the two triangles in a special way shown there, so this is *placement symmetry*. Potential symmetry is a kind of placement symmetry, occurring spontaneously in natural phenomena. Another kind of potential symmetry is latent symmetry.

But irrespective of whether we call it potential symmetry or placement symmetry, there are no surprises in Fig. 11.1(a). The rhombus has the symmetry $m_x m_y$, which is fully accounted for by the inherent symmetry m_y and the placement symmetry m_x. There is no latent symmetry involved.

By contrast, the symmetry of Fig. 11.1(b) is higher than that of the rhombus, and that is the surprise. The two component symmetries are still m_y and m_x only, and nothing else, and yet there is a 4-fold axis of symmetry emerging unexpectedly in the composite object, namely the square. This is a manifestation of latent symmetry, present in the right-angled isosceles triangle. Thus, latent symmetry is also a potential symmetry, except that its manifestation has a surprise element.

11.3 The fundamental theorem of symmetry

The BB in the above example of crystals of the second molecule is, by itself, devoid of 2-fold symmetry, or any other crystallographic symmetry. It is only the pair of the BBs, occurring in the unit cell of the crystal with space-group symmetry P2, that has 2-fold symmetry. We can generalize these considerations to arrive at an important theorem for crystallographic symmetry: *All translational and directional symmetry exhibited by a crystal is nothing but a self-organized manifestation of the potential symmetries residing in its constituent identical building blocks.*

This conclusion actually has a larger range of validity than just for crystallographic symmetry. Consider any object or system consisting of two or more subparts which are equal or equivalent. Any symmetry possessed by the object or system is because of the potential symmetry inherent in the subparts which becomes manifest symmetry when the subparts come together in a special or unique way to constitute the full object or system. We can therefore state what I call *the fundamental theorem of symmetry*:

> *Any spontaneously occurring symmetry of an object or system comprising of equal subparts is nothing but a self-organized manifestation of the potential symmetry residing in the subparts.*

11.4 The symmetry composition principle

Potential symmetry can be of two kinds: (i) placement symmetry; and/or (ii) latent symmetry. A geometrical example will help explain this (Fig. 11.2).

Fig. 11.2(a) is an isosceles triangle with $\theta \neq 90^0$. This is the building block (BB); let us call it component A.

Suppose we make a copy (m_xA) of the BB by reflecting it across the mirror plane (m_x) normal to the x-axis. Suppose further that we juxtapose the original BB and its copy in a special way shown in Fig. 11.2(b). We get a rhombus which has an additional mirror plane of symmetry, m_y. We call it placement symmetry (PS), because it has arisen because of the special placement of the two BBs. There is no surprise here.

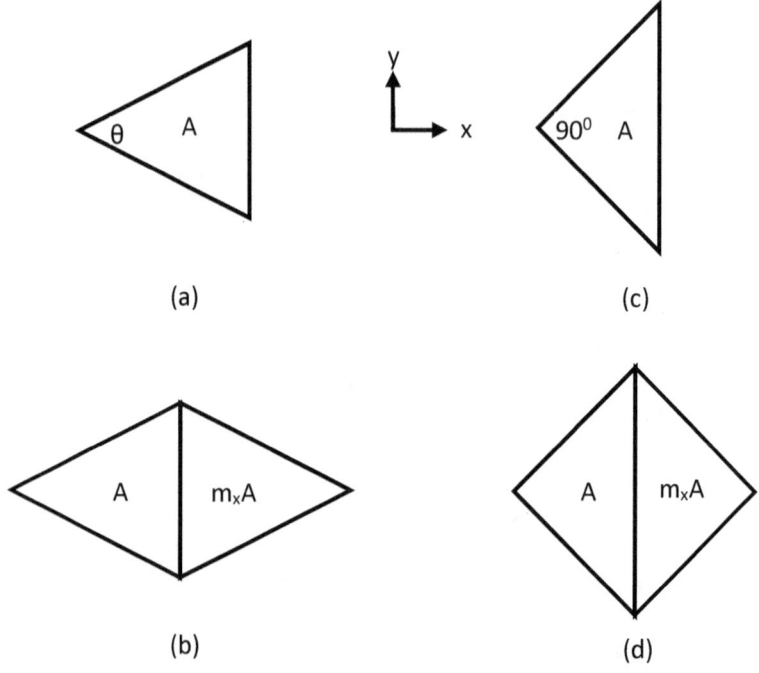

Fig. 11.2 Build-up of symmetry, starting from an isosceles triangle (a) or (c).

The surprise comes when we start all over again and build up the symmetry as before, except that now $\theta = 90^{o}$ for the starting triangle (Fig. 11.2(c)). The final result is Fig. 11.2(d), which is a square, and not a rhombus. It has a 4-fold axis of symmetry, which is a *latent symmetry* (LS). It is not a placement symmetry because the only placement symmetry here is m_y.

So we can generalize and come to an important conclusion, which I call *the symmetry composition principle*:

> *The existence of a symmetry often implies the coexistence of two or more equal or equivalent components or building blocks, and the overall symmetry group is then either the*

product of the symmetry group of the building block and the placement-symmetry group which describes the mutual placement of the building blocks, or it is a larger group because of the presence of latent symmetry.

Let H denote the symmetry group of the building block, and let G be the placement-symmetry group. The symmetry composition principle says that the symmetry group C of the overall (composite) object can be either equal to or higher than the direct product of H and G:

$$C \geq H \otimes G \tag{11.1}$$

When there is no latent symmetry present, $C = H \otimes G$.

The merit of this way of looking at symmetry is that it shifts the focus to the BBs. It is a *bottom-up* approach to symmetry, rather than the top-down approach in which we speak in terms of symmetry and broken symmetry. The bottom-up approach is about *emergent symmetry*, sometimes even unexpected emergent symmetry (if latent symmetry is involved).

Symmetry can be either a man-made contraption, or a natural phenomenon. In the manifestation of latent symmetry there is always the surprise element or the unexpected part in either case.

Placement symmetry, on the other hand, has no surprise element in either case. If it is a consequence of natural (as opposed to man-made) phenomena, then it is a case of *self-assembly* or *self-organization*, and is thus a manifestation of the potential symmetry of the BBs. Of course, for natural phenomena, manifestation of latent symmetry is also a case of self-organization.

In anticipation of the *partition theorem* for latent symmetry, which I shall describe in the next chapter, let us dwell some more on the reason for the presence of latent symmetry in the building block A, shown in Fig. 11.2(c), which manifested itself as a 4-fold axis of symmetry in the composite object shown in Fig. 11.2(d). As shown in Fig. 11.3 below, this latent symmetry was existing in a *subunit* of this object.

The BB can be divided into four equal subunits (which are right-angled isosceles triangles again), and pairs of these subunits are related by a 4-fold axis of symmetry. Thus in this example it is a symmetry lying latent in a *subunit* of the BB which becomes manifest when the composite object shown in Fig. 11.2(d), namely a square, is constructed from the BB and its copy. *Thus there can be symmetries of the subunits which are not symmetries of the BB, but become manifest in the symmetry of the composite object.*

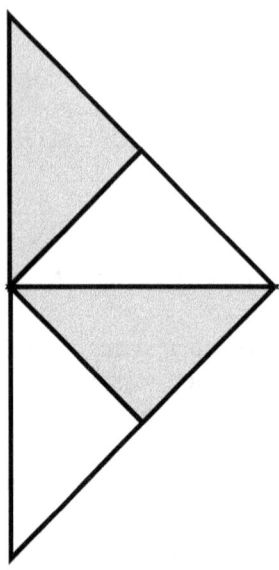

Fig. 11.3 Because the building block shown in Fig. 11.2(c) is a right-angled isosceles triangle, it can be divided into four equal subunits as shown here. It is seen that two of the subunits (shown shaded) are related by a 4-fold axis of symmetry, as are the other pair of subunits (unshaded). This 4-fold symmetry is present in portion of the building block, but not in the entire building block. It is this hidden symmetry which becomes symmetry when the composite object shown in Fig. 11.2(d), namely a square, is constructed from two right-angled isosceles triangles.

11.5 Placement symmetry

A symmetrical object has at least two equal parts, and its symmetry depends in a crucial way on the placement symmetry, i.e., on the way in which the equal parts are placed with respect to one another. Placement symmetry can either reduce (even destroy) or enhance (through the possible manifestation of latent symmetry) the overall symmetry of a composite object made up of equal parts.

Even when the equal parts have high symmetry, the overall symmetry can be as little as nil if the placement symmetry is nil.

On the other hand, the right kind of placement symmetry can make an object exhibit the full latent symmetry of the equal parts. In natural phenomena (as opposed to man-made contraptions), placement symmetry arises spontaneously through self-organization. Naturally, its emergence is compatible with the energy-minimization requirement. Thus, latent symmetry can manifest itself in natural phenomena only when conditions for its full or partial manifestation coincide with the least-free-energy requirement.

11.6 Latent symmetry and algorithmic information

Comprehension is compression.
Gregory Chaitin

Symmetry is a complexity-reducing concept (co-routines include subroutines); seek it everywhere.
Alan J. Perlis

Consider a sequence comprising of the set of all integers arranged in an ascending order. This set is infinitely large, so an infinitely large number of bits (0s and 1s) is needed for specifying it completely. We may tend to think that the set of integers has infinitely large information content. But there is something bothersome about that last statement. There is something systematic about the set of integers, and we can generate the entire set by writing a *small* computer algorithm (requiring only a small number of bits). We therefore say that the set of integers has only a small *algorithmic* information content (AIC). Thus the information content in the set of all integers can be *compressed* into a small number of bits needed to specify the algorithm for generating the entire sequence of integers. This becomes possible because we have *comprehended* the systematics of the way one integer can be obtained from the one preceding it in the sequence. Comprehension is compression.

Next consider a totally random sequence of integers. Because of the randomness, i.e., because of the lack of any systematics, the information is incompressible now. It is not possible to write an algorithm more compact (requiring a smaller number of bits) than the number of bits needed to write down directly the entire random sequence. The *'degree of complexity'* is very high in this case, and it is very low in the previous example.

Thus comprehension or knowledge about any systematics means a lower degree of complexity compared to a situation in which there is no such systematics or order present and comprehended. The comprehension of any *symmetry* therefore implies a compression (lowering) of the amount of algorithmic information needed to specify a system. Symmetry implies compression. Detection of symmetry means comprehension, or a lower degree of complexity. If symmetry decreases, degree of complexity increases, and *vice versa*.

When conditions are just right for latent symmetry to become manifest symmetry, the resultant (unexpected) increase in symmetry implies an extra increase in comprehension, or an extra decrease in the degree of complexity.

Pattern formation is a characteristic feature of complex systems (Wadhawan 2017). Patterns imply symmetry. Complex systems form patterns spontaneously, and the symmetry of the pattern may be either inherent symmetry plus placement symmetry, or inherent symmetry plus placement symmetry plus latent symmetry.

12. Group-Theoretical Determination of Latent Symmetry

Only a qualitative description of latent symmetry was given in the previous chapter. Any rigorous treatment of the subject requires the use of group theory. This was done by Litvin and Wadhawan (2001, 2002), and this chapter is based on that work.

12.1 Formal definition of latent symmetry

Consider an unordered set, S, of objects obtained by applying a set of isometries $\{g_1, g_2, ..., g_n\}$, with $g_1 = 1$, to an object A having intrinsic symmetry H. [An isometry is any distance-preserving mapping, or 'rigid motion'.] Thus

$$S = \{A, g_2 A, g_3 A, ..., g_n A\} \tag{12.1}$$

Let us *assume* that the set of isometries $\{g_1, g_2, ..., g_n\}$ form a group, and denote it by G:

$$G = \{g_1, g_2, ..., g_n\} \tag{12.2}$$

G thus is a representation of what I called placement symmetry in Chapter 11.

Latent symmetry was formally defined by Litvin and Wadhawan (2001, 2002) as any symmetry of the composite object that is not obtainable as a product of the isometries of H and G.

The 4-fold axis in Fig. 11.1(b) is an example of latent symmetry. For that case, $H = \{1, m_y\}$ and $G = \{1, m_x\}$, so the composite object is expected to have only the symmetry $C = \{1, m_x, m_y, 2_z\}$, unless there is some latent symmetry involved. The unexpected 4-fold axis of symmetry is a latent symmetry which became manifest only when the composite object was created, or got created, in the special way shown in Fig. 11.1(b).

Another example of an unexpected symmetry jump because of the presence of latent symmetry has been given by Flack (2003).

12.2 Litvin's partition theorem for latent symmetry

The partition theorem for latent symmetry was proved and enunciated in
Litvin and Wadhawan (2002). Given a composite S defined by Eq. 12.1 in
terms of the isometries of the group G (defined by Eq. 12.2) applied on a
component A, the partition theorem is:

> *If the component A can be partitioned into a set of
> symmetry-related subunits $\{B, v_2 B,..., v_m B\}$, related by a
> set of isometries $\{1, v_2,..., v_m\}$ which are a set of right-
> coset representatives of a group V in a right-coset
> decomposition of V with respect to G, then V is an
> invariance group of the composite.*

It should be noted that the theorem states that V is *an invariance group* of
the composite, and not necessarily *the symmetry group* of the composite. An
invariance group is a group comprising of *any* set of isometries which leave
the composite invariant. The symmetry group is the set of *all* isometries that
leave the composite invariant.

Latent symmetry is a symmetry of a subunit of the basic component A that
is also a symmetry of the composite. In general, latent symmetry does not
necessarily give us the entire symmetry of the composite (see Example 2
below).

Example 1

Fig. 12.1(a) shows a component A having intrinsic symmetry $H = 1$. Using
a group of operations $G = \{1, m_1\}$, a composite S is constructed from A:
$S = \{A, m_1 A\}$. It has pentagonal symmetry.

Fig. 12.2(a) shows how the component A can be partitioned into five
subunits:

$$A = \{B, m_2 B, 5_z B, m_3 B, 5_z^{\,2} B\} . \tag{12.3}$$

The isometries here are defined in Fig. 12.2(b). The z-axis is perpendicular
to the plane of the paper.

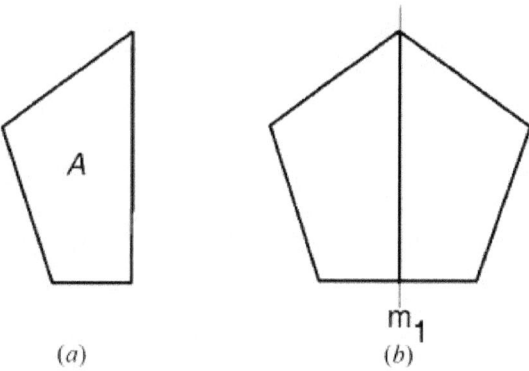

Fig. 12.1 Component A (a), and a composite constructed from it by juxtaposing a mirror image (b). [Litvin and Wadhawan (2002).]

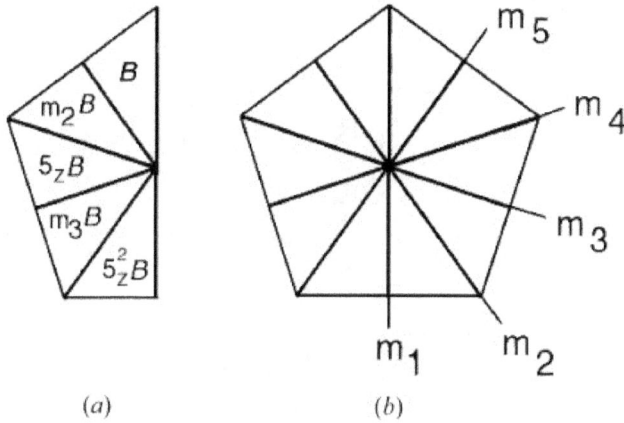

Fig. 12.2 (a) Partitioning of the component A into symmetry-related subunits. (b) The isometries used for the partitioning. [Litvin and Wadhawan (2002).]

A group $V = 5_z mm$ can be identified such that the set of isometries $\{1, m_2, 5_z, m_3, 5_z^2\}$ constitutes a set of right-coset representatives of the coset decomposition of $5_z mm$ with respect to m_1. Therefore the composite object is invariant under the group $5_z mm$.

Example 2

The shaded volume in Fig. 12.3 shows an isosceles triangular prism, which is the component A. It has the intrinsic symmetry $H = m_z$. Fig. 12.3 also shows the full cube that can be constructed from this component by the isometries of the group $G = 4_z m_x m_{xy}$.

Fig. 12.4 shows a partitioning of the component, namely the shaded volume in Fig. 12.3 (shown again as Fig. 12.4(a)) into six symmetry-related subunits, obtained by the set of isometries $\{1, m_{\overline{xz}}, 3_{xyz}, m_z, 4_y, 3-\overset{5}{\underset{xyz}{}}\}$.

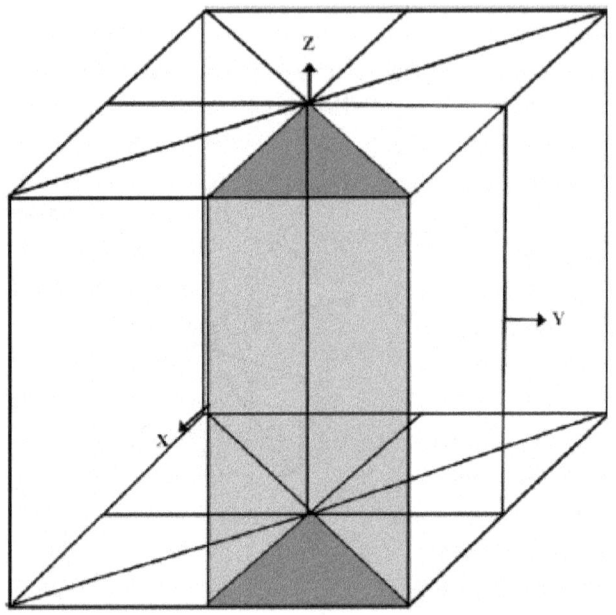

Fig. 12.3 A component of symmetry m_z (shaded area), and the composite object (full cube) constructed from it by the operations of the group $4_z m_x m_{xy}$.
[Litvin and Wadhawan (2002).]

This set is a set of right-coset representatives of the decomposition of the group $V = m\overline{3}m$ with respect to the group $G = 4_z m_x m_{xy}$. Thus the composite is invariant under the operations of the group V.

A composite object exhibiting symmetry may be either man-made, or of natural occurrence. If it is man-made, then we have knowledge of the isometries G used for constructing the composite from a component A whose symmetry H is also known. When G and H are known, as also the structure of A, the partition theorem can be helpful in detecting the latent symmetry which became manifest when the composite was constructed.

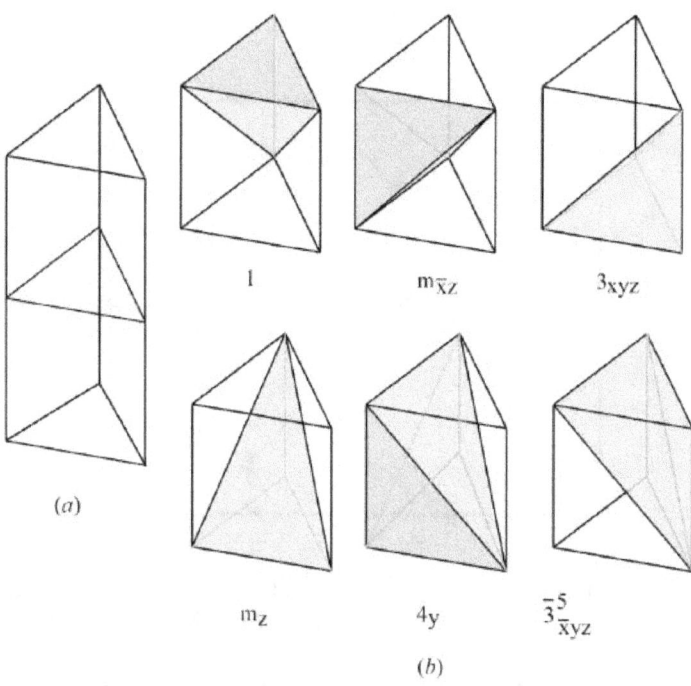

1 $m_{\bar{x}z}$ 3_{xyz}

(a)

m_z $4y$ $3\frac{5}{\bar{x}yz}$

(b)

Fig. 12.4 (a) A component divided into upper and lower halves by its intrinsic symmetry m_z. (b) Partitioning of the component into six symmetry-related subunits. The top three subunits are for the upper half of the component, and the bottom three for the lower half. Shown below each subunit is the isometry used for obtaining it from the first subunit on the top left. [Litvin and Wadhawan (2002).]

<u>Converse of Litvin's partition theorem is not true</u>

The converse of Litvin's partition theorem is not true (Litvin and Wadhawan 2002). Fig. 12.5 illustrates this. It shows a composite of symmetry $4_z m_x m_{xy}$ constructed from a component having no intrinsic symmetry (i.e. $H = 1$).

In this example, $G = 4_z$. No possible right-coset representative v_2 of the coset decomposition of $4_z m_x m_{xy}$ with respect to 4_z, namely any of the operations $m_x, m_y, m_{xy}, m_{\bar{x}y}$, can be used for partitioning the component

A into two subunits $\{B, v_2 B\}$.

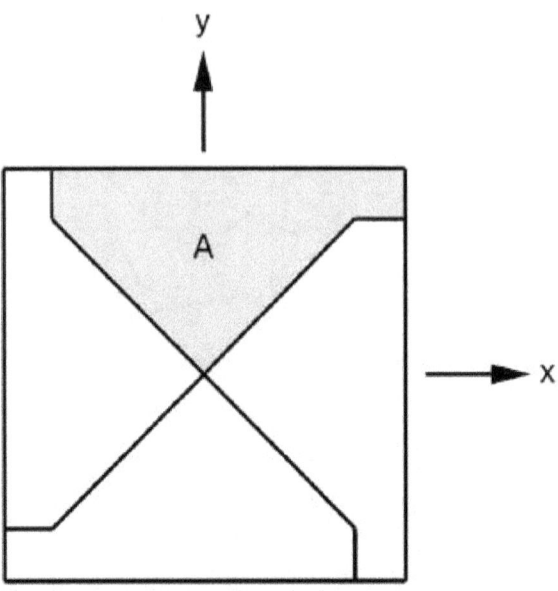

Fig. 12.5. Shown here is a square, constructed from a component A (shaded area) which has no intrinsic symmetry. [Litvin and Wadhawan (2002).]

12.3 Latent symmetry and domain-average engineered ferroic materials

At a ferroic phase transition from a prototype symmetry G to a lower symmetry F ($F \subset G$), the ferroic phase can exist in a number of domain states or orientation states. This number (n) of domain states is given by the ratio of the orders of the groups G and F. These domain states correspond to the n cosets in the following coset decomposition (see Wadhawan 2000):

$$G = F + g_2 F + ... + g_n F \qquad (12.4)$$

In an ideal unconstrained and unbiased situation, the domains are distributed randomly in the sample, and each domain type occupies a fraction $1/n$ of the volume. Therefore the average properties of the ferroic specimen tend to mimic the situation as if no phase transition has occurred. For example, if the ferroic phase is piezoelectric and the prototypic phase is centrosymmetric, the effective piezoelectric coefficients will be zero because the piezoelectric responses of the various domains will get cancelled out. Attempts are therefore made (by a process called 'poling') for introducing a frozen-in bias in the domain structure so that some domain types are favoured over the others, thus creating a nonzero average piezoelectric response (Wadhawan 2000).

Such considerations have been generalized and symmetry analyses have been carried out for situations in which the actual number of domain types (say m) in a given ferroic specimen is deliberately made less than n (Fousek and Cross 2001; Fousek, Litvin and Cross 2001). So, what we get is an *engineered* ferroic: a so-called domain-average engineered (DAE) ferroic.

A ferroic phase, whether engineered or not, is a conglomeration of a certain large number of randomly distributed domains, each with symmetry D_i, with

$$D_i = g_i D_1, \qquad i = 1, 2, \dots n \tag{12.5}$$

where the symmetry of domain type D_1 is F.

Similarly the symmetry of a DAE sample is the symmetry of the superposition set given by

$$\{D_1, D_2, \dots D_m\} = \{D_1, g_2 D_1, \dots g_m D_1\} \tag{12.6}$$

It was pointed out by Litvin, Wadhawan and Hatch (2003) that the domains in an ideal specimen are equal or equivalent objects, and therefore the possibility of unexpected manifestation of latent symmetry can arise. Fig. 12.6 provides an illustration of this.

Fig. 12.6(a) shows a 2-dimensional crystal A having symmetry $F = p2_z$. Fig. 12.6(b) is a composite obtained by superposing A with its copy $g_2 A$, with $g_2 = m_y$. And its symmetry is $p4_z m_x m_{xy}$, with no change in translational symmetry. This superposition of the two domain types has the symmetry elements $4_z, 4_z^{-1}, m_{xy}, m_{\overline{xy}}$, which are not products of the elements of $F = p2$ and the elements of $\{1, m_y\}$. Therefore these are manifestations of latent symmetry.

12.4 An example of how ignorance about latent symmetry can lead to errors

Errors can arise if one is not aware of the possibility of unexpected manifestations of latent symmetry. Here is an example from the work of Vlachavas (1984), as discussed in Wadhawan (1987) and Litvin, Wadhawan and Hatch (2003). Vlachavas proved two theorems:

<u>Theorem 1</u>. Given a two-component composite $\{A, gA\}$, where the component A is of point-group symmetry F. The order of the point-group symmetry of the composite is $2/k$ times the order of the group F, where k is a positive integer.

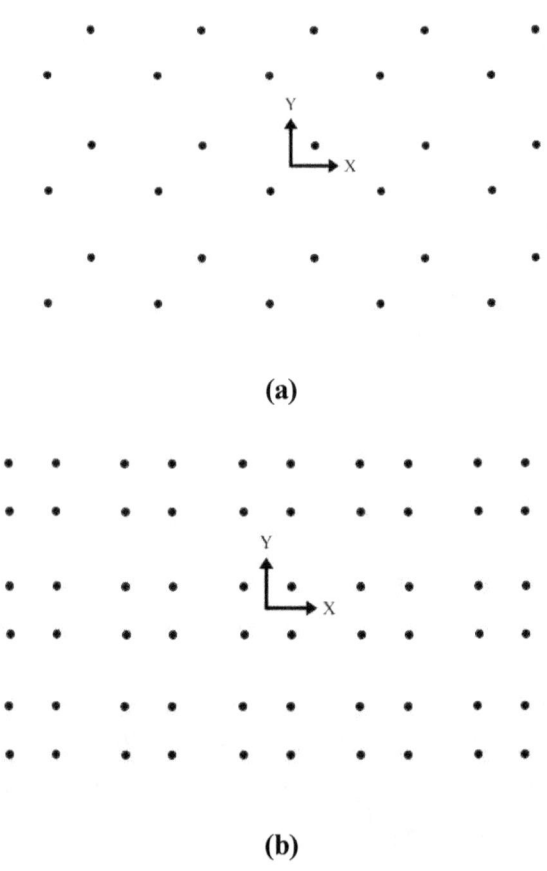

(a)

(b)

Fig. 12.6 (a) A 2-dimensional crystal; let us call it component A. (b) The composite obtained by the superposition of A with $m_y A$.
[Litvin, Wadhawan and Hatch (2003).]

<u>Theorem 2 (a corollary of Theorem 1)</u>. The lowest order of the composite point-group symmetry is 2, and the highest is 2 times the order of the group F.

These theorems cannot be valid in certain latent-symmetry situations. Fig. 12.6 is an example. For it, $F = 2_z$; i.e., it is a point-group of order 2. The theorems predict that the order of the symmetry group of the composite shown in Fig. 12.6(b) can be only 2 or 4. But the actual point group here is $4_z m_x m_{xy}$, a group of order 8.

Another example of the violation of these theorems is provided by Fig. 11.1. Fig. 11.1(b) (for $\theta = 90^0$) is shown again in Fig. 12.7.

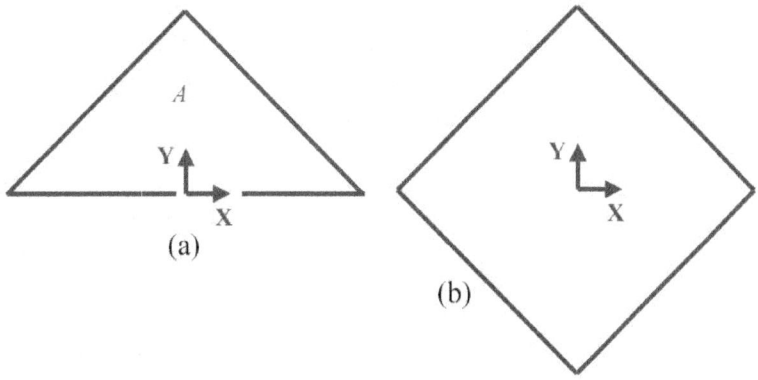

Fig. 12.7 A 2-dimensional component A. (b) The composite $S = \{A, m_y A\}$.
[Litvin, Wadhawan and Hatch (2003).]

For this case, $F = \{1, m_x\}$, i.e. a group of order 2. The composite constructed from it by the isometries $\{1, m_y\}$ has the symmetry $G_s = 4_z m_x m_{xy}$, a group of order 8. But Vlachavas's theorems wrongly predict this order to be either $(2/1) \times 2 = 4$ or $(2/2) \times 2 = 2$.

This violation of the theorems occurs because there is a kind of 'singularity' in the variation of the order of the group G_s with θ at $\theta = 90^0$ (see Fig. 12.8). The order is 4 all through, except when $\theta = 90^0$; it takes the value 8 then.

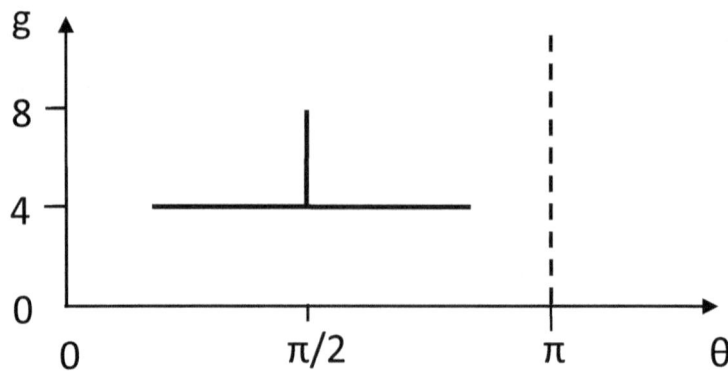

Fig. 12. 8 The order (g) of the point group of the composite shown in Fig. 11.1(a) is 4 for all values of θ except for $\theta = 90^0$. It is as if there is a 'singularity' for this value of θ. [Wadhawan (2000).]

This is what latent symmetry is all about. As some parameter changes continuously, suddenly there is a special value of the parameter for which the symmetry of the composite system is higher. And it can happen unexpectedly.

Vlachavas made the mistake of assuming that there are only two types of symmetry that leave invariant a composite made up of two components 1 and 2:

(i) Symmetries that map each component back onto itself.

(ii) Symmetries that map Component 1 to Component 2, and Component 2 to Component 1.

But actually a third kind is also possible:

(iii) Symmetries that send a *subunit* of Component 1 to a subunit of Component 2, and the rest of Component 1 to another subunit of Component 2. And *vice versa* for Component 2.

12.5 The role of placement symmetry in revealing latent symmetry

The manifest symmetry S of any composite object made up of equal building blocks (BBs) depends on the placement symmetry G and also on any latent symmetry in the building block. The expected effect of the placement-symmetry part is straightforward; there are no surprises. But any latent symmetry present and revealed depends on what the placement symmetry happens to be. One extreme situation is that in which the set of copies of the BB is just some jumbled collection, so that all information about any possible latent symmetry is lost, or not revealed. At the other extreme, one may end up having a placement symmetry that reveals the full latent symmetry of the BB.

In this context it is relevant to mention here the well-known notions of '*invariance group*' (mentioned earlier) and '*partial symmetry*'.

Consider a composite object made up of two or more identical BBs. Its symmetry group is that which consists of *all* isometries that map it back onto itself. But it can also have a number of invariance groups: By definition, an invariance group is *any* group of isometries that leave the composite object invariant, but not necessarily the group of *all* such isometries. So there is only one symmetry group, but possibly many invariance groups.

Similarly, there is a distinction between *global symmetry* and *partial symmetry*. It can happen that two subunits of an object are related by an isometry, but the object as a whole is not mapped back onto itself by this isometry. Thus it is a partial symmetry but not a global symmetry of the object. Partial symmetry is related to concepts such as groupoids, semi-groups, etc. In crystallography such ideas have been applied to order-disorder crystal structures or *OD crystal structures* like polytypes (see Fichtner 1986).

The latent symmetry of a given composite object (if at all the BB of the object has any latent symmetry) manifests itself through the operations of the placement symmetry group G. And G may well comprise of partial-symmetry operations. Latent symmetry can be a symmetry of a *subunit* of the BB (but not of the entire BB) which is also a symmetry of the composite object. There are examples of this in Fig. 11.3 and Fig. 12.4.

There is a clear difference between latent symmetry and partial symmetry, even though both can be symmetries of a subunit of the BB. Latent symmetry is *always* a part of the overall symmetry of the composite object. By contrast, a partial symmetry is a non-space-group symmetry (in the case of crystals); it is not manifest in the space-group symbol.

12.6 Concluding remarks

I have focussed only on geometrical examples for explaining the idea of latent symmetry. Clearly the idea has a general validity, independent of the type of group needed for describing the symmetry. Only crystallographic groups were considered so far. We shall encounter *permutation groups* in the next chapter when I discuss the symmetry of complex networks.

The idea of latent symmetry is relevant not only for objects, but also for phenomena, fields, networks, etc. It is conceivable, even likely, that certain symmetry jumps in natural phenomena (e.g. phase transitions in crystals) may well be manifestations of latent symmetry occurring under certain *just-right* conditions. This needs to be looked for, and investigated.

13. Symmetry of Complex Networks

Symmetry and complexity determine the spirit of nonlinear science. The expansion of the universe, the evolution of life and the globalization of human economies and societies lead from symmetry and simplicity to complexity and diversity.

Klaus Mainzer, *Symmetry and Complexity*

A complex system usually consists of a large number of simple 'members', 'elements', or 'agents', which are interdependent and which have the potential to generate qualitatively new collective behaviour. The manifestations of this behaviour are the spontaneous creation of new spatial, temporal, or functional structures (Wadhawan 2010, 2017).

Network theory provides a powerful approach for investigating complex systems. Each member or element of the system defines a node, and specific pairs of nodes are taken as connected by an edge if there is an interaction between that pair. A complex system can be thus regarded as a *complex network* (CN) (Barabási 2003, 2009).

Several real-life networks are rich in symmetry, and exploring the origins of that symmetry can provide insights useful for modelling the dynamics and topology of the network (MacArthur and Anderson 2006; MacArthur, Sánchez-García and Anderson 2007; Holme 2006; Xiao *et al.* 2008a, b).

13.1 Latent symmetry in complex networks

. . similar linkage pattern is the underlying ingredient that is responsible for the emergence of symmetry in complex networks.

Xiao *et al.* (2008b)

A characteristic feature of complex systems is the emergence of *unexpected* or *unpredictable* properties or behaviour (Wadhawan 2010, 2017). As we have seen in this book, unexpected symmetry-jumps in assemblies (composites) comprising of equal or equivalent objects or phenomena is also a feature of latent symmetry. One is tempted to speculate that there can be a connection between the two in certain situations.

The existence of symmetry in complex networks has been investigated relatively recently (see Xiao 2008a, b). Studies have shown that *most of the real-life networks possess a high degree of symmetry* (MacArthur and Anderson 2006; MacArthur, Sánchez-García and Anderson 2007). This is reminiscent of the fact that most biological systems possess symmetry of some kind, with evolutionary underpinnings. The so-called '*similar linkage patterns*' have been observed in many networks. The phrase 'similar linkage patterns' means that vertices with the same degree tend to share similar neighbours.

The word 'emergence' is used extensively when describing the essence of complexity. A complex system has emergent properties; i.e., it has properties that we cannot always expect or predict from the properties of its constituents or agents. From our discussion of latent symmetry in this book, we can say that if a complex system exhibits an *increase* of symmetry (rather than a decrease of symmetry normally associated with increase of complexity, as typified by Klaus Mainzer's statement at the top of this chapter), the increase *may* sometimes imply the presence of equivalent subunits which possess latent symmetry. This latent symmetry can lead to emergent symmetry when the subunits come together or interact in a special way necessary for the manifestation of the latent symmetry.

It should be possible to extend the notion of latent symmetry from crystals to other systems consisting of equal, identical or equivalent objects or agents. Such systems, for example complex networks, may have at least the exchange symmetry or permutation symmetry.

There are two important features of latent symmetry and its manifestation (emergent symmetry) which make it relevant to an understanding of a large class of complex systems:

1. Only certain special mutual configurations and linkages in a complex system of identical or equal agents can make manifest the full latent symmetry of the agents.

2. When the conditions are conducive for the full manifestation of latent symmetry, i.e., when a system has hit upon or discovered the special mutual configurations mentioned above, there would be an *unexpected* increase of symmetry. The corresponding unexpected change of order is in line with the crux of complexity, namely the *unexpected* emergence of new features or properties.

Thus, latent symmetry and its manifestations *may* be playing important roles in the self-assembly and self-organization of complex networks, leading to possible evolutionary or other advantages. We have already seen in Section 9.9 Dawkins' (1996) discussion of the restrictions imposed by symmetry in the context of Darwinian evolution, and how such constraints can influence the evolvability of certain life forms.

Symmetry analyses of networks have led to the important result that similar linkage patterns are the underlying factor responsible for the manifestation of symmetry by networks (Xiao *et al.* 2008b). This similar-linkage-pattern idea can be linked to the latent-symmetry idea, at least in some situations. How?

To answer this question I reproduce here something I wrote in Section 11.1:

'What decides whether the crystal grows to have symmetry P2, and not P1? Obviously, it has to do with the shape of the molecule in the building block (BB), and the nature of the force field around it which is responsible for the interactions among the various copies of the BB in the fluid from which the crystal has grown, unit by unit. If the symmetry of the bulk crystal is P1, it means that, through a process of trial and error, the set of identical molecules (equal objects) constituting the crystal has found a least-free-energy configuration such that the assembly has translational symmetry, but no directional or point-group symmetry.

'If the symmetry of the crystal is higher, say P2, it means that the shape and the governing interactions among the identical molecules (the BBs) are such that the least-free-energy configuration of the resultant crystal is one in which there are *two* BBs in the unit cell, *and which are so oriented with respect to each other* that we can obtain the coordinates of the atoms in one BB by a 2-fold rotation of the other BB about a suitably located axis of symmetry.

'It is important to realize that this 2-fold symmetry of the crystal is entirely a consequence of tendencies and forces present in the BB or the molecule. The shape of the molecule is such, and the nature of the forces emanating from it are such, that we can say that 2-fold symmetry was inherently present as a *potential* symmetry (*latent or/and placement symmetry*), and became manifest when two such identical molecules got a chance to interact with one another repeatedly till they could together settle down to a unique energy-minimizing mutual orientation and disposition with a manifest 2-fold symmetry.'

For exactly the same reasons, the 'similar linkage patterns' are the building blocks of real-life symmetric complex networks. When some subunit of a network has discovered (or has evolved to have) a certain linkage pattern, it stands to reason that other subunits of the network will also hit upon the same linkage pattern. For example, persons with similar background tend to flock together. People with similar educational background, interest, or age are likely to have common friends. If some such linkage pattern exists in some parts of society, the same may occur in other parts also. This is just like in crystal growth. If some parts of the growing crystal settle for an optimum linkage pattern, or growth with preferential attachment, other parts will also discover the same optimum linkage pattern or chemical bonding, corresponding to a specific symmetry.

13.2 Measures of symmetry of networks

Measures of exact symmetry

If a graph is shown to be symmetric, the next question is: How symmetric is it? I describe two measures of such symmetry here for deterministically created artificial networks (Xiao et al. 2008a, b).

The order $|\text{Aut}(G)|$ or α_G of the automorphism group of a graph is clearly a measure of the extent of symmetry in the graph. We should normalize it to be able to compare the symmetries of graphs of different orders N. The following is one such normalized measure of symmetry (MacArthur and Anderson 2006):

$$\beta_G = (\alpha_G / N!)^{1/N} \tag{13.1}$$

Fig. 4.1 and Eq. 4.9 indicate that a network or graph with more nontrivial orbits is more symmetric. Xiao et al. (2008b) have used a measure of symmetry which accounts for this:

$$\gamma_G = \frac{\sum\limits_{1 \le i \le k, |V_i| \ge 1} |V_i|}{N} \tag{13.2}$$

Here V_i is the ith orbit in the automorphism partition, and k is the number of cells in the partition.

Measures of approximate symmetry

Real-life networks are prone to a certain degree of randomness, and therefore only their stochastic symmetry (if any) is of practical relevance. Parameters defining stochastic or statistical symmetry are of a *continuous* nature: Instead of giving a yes/no answer to the question of whether a symmetry is present, one speaks in terms of the *extent* of symmetry present in an average sense. Holme (2006a, b) introduced the notion of *degree symmetry* in this context. The degree of a vertex in a network is the number of other vertices linked to it. Degree symmetry is a measure of the number of paths (i.e., the non-self-intersecting sequences of connected links or edges) going out from a vertex that have the same (or overlapping) degree sequences.

Holmes (2006a) arrived at a measure of the degree symmetry of a vertex by performing walks along all paths from the vertex and comparing the sequence of degrees of the vertices along the paths. Consider the vertex shown as a triangle in Fig. 13.1. All walks from it pass through vertices which have degrees 3, 2, ... Therefore this vertex has a very high degree symmetry.

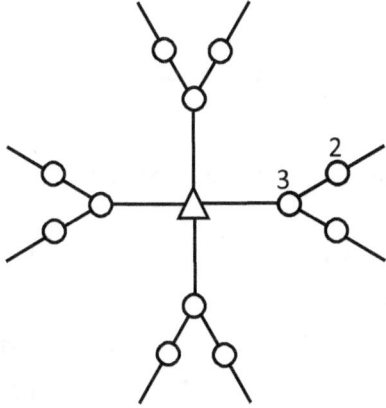

Fig. 13.1 The central vertex in this network has a very high degree symmetry, because all paths from it pass through vertices which always have the degree sequence 3, 2, . . [After Holmes (2006a).]

Identification of such a symmetry in a real network leads to the next question: What mechanisms or tendencies were responsible for this high degree-symmetry? And the question is relevant even when the observed symmetry is only approximate, and not exact.

Holmes (2006a, b) defined a certain quantitative *degree coefficient* in this context, and applied it to some real networks. He pointed out that the presence of, for example, network motifs in biological networks is a pointer to the relationship between degree symmetry and functionality.

13.3 Origins of symmetry in complex networks

> *The ubiquity of symmetry in disparate real-world systems suggests that it may be related to generic self-organizational principles.*
>
> MacArthur and Anderson (2006b)

It is my assertion that symmetry in real-life CNs is a secondary organizing principle, just as it is a secondary organizing principle for crystals (see Section 5.3). Similar linkage patterns are involved in both cases. Symmetry in a CN is an indicator and a measure of the organizing forces responsible for creating the network that departs significantly from the corresponding random network having the same degree-distribution etc. (called the *null model*). This conclusion can be drawn from the work of Xiao *et al.* (2008b). I summarize their work in this section.

Xiao *et al.* (2008b) analysed the statistical data about the symmetry of a variety of complex networks, and concluded that in all cases there exist what they formally defined as similar linkage patterns or SLPs. The SLPs can be understood and represented in terms of the so-called *symmetric bicliques*. Fig. 13.2 shows a couple of examples of symmetric bicliques.

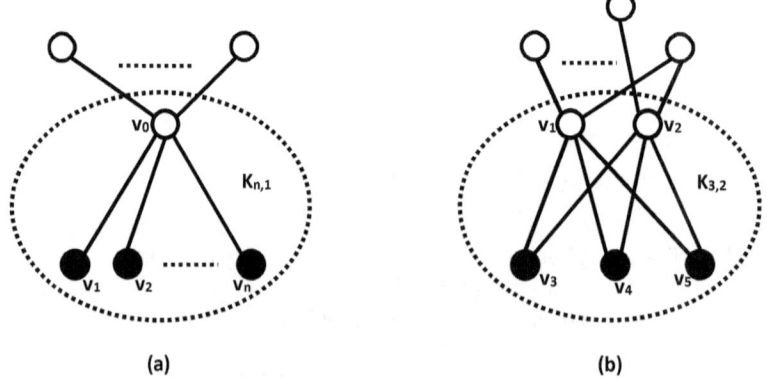

(a)   (b)

Fig. 13.2 (a) A symmetric biclique, $K_{n,1}$ which contributes to the size of the automorphism group of the network with a factor n!. (b) A symmetric biclique, $K_{3,2}$ which contributes to the size of the automorphism group with a factor 3!. [After Xiao *et al.* (2008b).]

Consider two disjoint vertex sets V_1 and V_2, with the property that vertices in the same set are not connected (i.e., are not adjacent), and all vertices in a set are connected to all vertices in the other set. Fig. 13.2(a) is an example of this, with $V_1 = v_o$ and $V_2 = (v_1, v_2, ... v_n)$. In such a case, the combined set of all these vertices (enclosed inside the dotted-line oval in Fig. 13.2(a)) is a complete bipartite graph, denoted by $K_{v1,v2}$ (or $K_{n,1}$ for the example of Fig. 13.2(a)).

Further, if $K_{v1,v2}$ is a subset of a larger graph $G(V, E)$, and for every vertex v in V_1 the neighbour set of v in the graph K is the same as the neighbour set of v in the graph G, then $K_{v1,v2}$ is called a *symmetric biclique* (Xiao *et al.* 2008b).

It follows from this definition that in such bicliques, vertices in V_1 have the same degree and share the same neighbours. This is therefore an example of a graph with a similar linkage pattern (SLP).

Fig. 13.2(b) provides another example of an SLP, with $V_1 = (v_1, v_2)$, and $V_2 = (v_3, v_4, v_5)$.

From Fig. 13.2 one can see the analogy between chemical bonding in a crystal and adjacency or connections in a complex network. The degree of a vertex is like the number of bonds around an atom as decided by its valence (with a double bond counted as two bonds, etc.). And sharing of same neighbours in a symmetric biclique is like the chemical bonding to the same atomic species in the structure of a crystal. Therefore, social bonding in a social network has its analogy with chemical bonding in the atomic structure of a crystal, with symmetry in both cases being determined by SLPs. Thus the observed symmetry has similar origins in the two cases.

13.4 The similar-linkage-pattern model for symmetry

Xiao *et al.* (2008b) have proposed an important model for the symmetry observed in complex networks, which I outline here.

This model starts off from *the BA model* I mentioned in Section 6.5, with Eq. 6.1 describing the power-law distribution of the degrees of the vertices. The BA model assumes *growth* of the network, and *preferential attachment* in the way new vertices attach to existing vertices. A highly attached existing vertex is more likely to get bonded to the new vertex. Let m denote the existing neighbours of a vertex. Xiao *et al.* (2008b) call this the *initial degree* of the vertex. Thus the new vertex preferentially attaches to m existing highly connected vertices. The 'initial degree' is a property associated with a vertex, and is akin to the 'fitness' or 'hidden variable' notion in complex networks (Caldarelli *et al*. 2002; Soderberg 2002).

In the BA model, m was taken as a constant. Xiao *et al*. (2008b) have shown in the model proposed by them that not treating m as a constant is an important step for reproducing the emergence of symmetry in real-life networks. And another change they introduce in the original BA model is the incorporation of similar linkage patterns (SLPs). Their model therefore has the following two ingredients:

(i) New vertices are assumed to get linked to existing vertices, not only obeying the principle of preferential attachment, but also that of SLP. Thus, the newly added vertices with initial degree m tend to link to targets to which existing vertices with degree m are linked.

(ii) The initial degree m of newly added vertices follows a certain distribution law, rather than being a constant.

Extensive simulation work by Xiao *et al*. (2008b) led to the conclusion that SLP is responsible for the emergence of symmetry in complex networks. They found that solely preferential attachment, with the initial degree following a distribution, does not necessarily reproduce the observed symmetry of networks. Preferential attachment alone, with a distribution for the initial degree, can only reproduce treelike symmetry in some cases.

The fact that there is a statistically significant occurrence of symmetric bicliques in real networks points to the existence of SLPs as the underlying organizational principle.

13.5 The free-energy landscape for biological networks

> *Exact symmetry in biology would even seem to be antithetical to the notions of complexity, variety, and metamorphosis that are central to the idea of life as we know it. Nevertheless, as in microphysics, life requires stability and sameness as well as change. The apparent conflict of these notions was captured in Schrodinger's metaphor of life as an "aperiodic crystal".*
>
> P. G. Wolynes (1996)

The analogy of complex network symmetry with crystalline symmetry raises a fundamental question. Symmetry arises in a crystal for satisfying the tendency to minimize the potential energy or the binding energy of the crystal. A jumbled-up conglomeration of the molecules would have a higher potential energy than that of a crystalline configuration which is inherently a close-packed configurations. What is the analogue of potential energy (binding energy or cohesive energy) for the case of a CN? To answer this question we should consider some specific CNs.

Symmetry and the folding efficiency of proteins

> *Most modern researchers see symmetry as an emergent feature of the general pasimony of our observed universe, resulting from the limited modes of interaction between a small number of building blocks as they assemble (or are assembled) into structures of greater complexity.*
>
> Goodsell and Orsol (2000)

A protein is a sequence of amino acids. This sequence defines the so-called *primary structure* of the protein. There is a backbone chain of the building blocks, with side chains (distinctive of the particular amino acids involved) attached to the backbone. Chemically, a large number of single bonds are present, so the long macromolecule can rotate along the single bonds, this freedom offering a huge number of 'folded configurations' which define the so-called *secondary structure* of the protein. Predicting protein folding from the primary structure is one of the great challenging problems of modern biophysics.

The problem is akin to (but much bigger than) that of crystal growth. Given a particular chemical species, and a specific solvent, can we predict the structure and symmetry of the crystal that would grow from the solution in a given set of ambient conditions? There are an infinite number of ways in which a protein can fold. Can we predict the protein folding? Hardly.

But a common factor in both crystal growth and protein folding is that both opt for a high degree of symmetry (Wolynes 1996; Li *et al*. 1996; Goodsell and Olson 2000; Levy *et al*. 2005; Andre *et al*. 2008). The explanation in the case of crystal growth is the one given above in this chapter, namely a tendency to have maximum possible binding energy. The same kind of explanation can be expected to hold for protein folding also. Researchers have found that this is indeed the case (see, e.g., Andre *et al*. 2008). Folded proteins tend to have a 'funnelled' free-energy landscape (Wolynes 1996;

Levy *et al.* 2005).

This fact (preponderance of high binding-energy configurations) can be rationalized even from the evolutionary-fitness viewpoint. High binding energy (as the cause of protein symmetry) means greater stability, thus increasing the chances of the occurrence of such configurations in larger numbers in the population. This, in turn, bestows an evolutionary advantage to the high-symmetry structure (Andre *et al.* 2008).

13.6 Social networks and the meaning of cohesive energy

The idea of high binding energy or cohesive energy leading to symmetry can be carried over to social networks also. The observation of *degree symmetry* (Holme 2006a, b), *growth with preferential attachment* (MacArthur and Anderson 2006), and *similar linkage patterns* (Xiao *et al.* 2008b) in complex networks indicates the same basic mechanism at work. The social, cultural, and biological networks acquire cohesiveness and symmetry in a concomitant manner. For example, people with similar interests and backgrounds are found to have common friends, thus acquiring a social cohesiveness (Castellano, Fortunato and Loreto 2007). The equivalent of binding energy here would be the effort or energy needed to break this social cohesiveness.

Energy is defined in physics as the capacity for doing mechanical work. Its most notable feature is that it is conserved in a wide variety of situations. If we regard this as the essential feature of energy, the same term can be used in other, non-physics, contexts as well (for example in neural-network dynamics). The conserved nature of energy constrains and often completely determines the dynamical and statistical behaviour of a system, in physics as well as elsewhere. *An even more general definition of energy is as something which drives change.* This last definition (though prone to misuse) is relevant for application to social networks.

Eq. 6.6 in Chapter 6 defines the Hamiltonian and the total energy of a complex network. The same section in Chapter 6 gives the expressions for entropy, partition function etc. Naturally, free energy can be defined, and the second law for open systems can be invoked. It is clear that the second law is the primary organizing principle resulting in the emergence of symmetry, not only in crystals, but also in the internet, the international trade network (Ruzzenenti *et al.* 2010), the international production network, and a variety of biological, neurological, and social networks.

In every case, symmetry emerges, and often equal (or near-equal) placement of equal (or near-equal) parts occurs, for maximizing the cohesive energy (as in crystals) or social cohesion (as in social networks).

14. Afterword

Symmetry signifies invariance to certain changes. A symmetric system often has two or more equal or equivalent subparts, which are related by certain operations or transformations (called symmetry operations or symmetry transformations).

At extremely high temperatures or energies, we have the highest symmetry, where even the four fundamental interactions of Nature merge into one. This is a state of maximum disorder. At lower temperatures, symmetry-breaking transitions occur. The lower temperatures enable the ordering tendencies to overcome some of the disorder characteristic of higher temperatures. By investigating the consequences of the broken or lowered symmetry, i.e., by examining the lower-symmetry phases or configurations, we can infer a lot about the nature of the interactions that drove the phase transition. This is the *top-down* approach to symmetry.

The *bottom-up* approach is more difficult, and therefore not very well developed at present. But the great advantage of the bottom-up approach to symmetry build-up is that it can help reveal the underlying organizing principles or symmetrizing principles. This approach helps answer questions like the following: Why there is so much symmetry around us? Why is it that, very often, molecules having no symmetry at all form crystals which have not only translational symmetry, but also directional symmetry?

Of course, the mother of all organizing principles is the generalized version of the second law of thermodynamics, applicable to open systems, i.e., systems which can exchange energy and/or matter with the surroundings. A phenomenon can occur only if the second law allows it to occur. And the second law says that, for all natural phenomena, $\Delta F \le 0$, where $F = E - TS$ is the free energy.

The entropy term TS is a measure of disorder. Order and organization can arise if the conditions are such that the decrease in the internal-energy term E is larger than the decrease in the magnitude of the entropy term TS. Growth of a crystal from a melt is a familiar example of this. The fluid state prevailing at higher temperatures is more disordered and has higher entropy. As the temperature is lowered, the crystalline phase becomes possible because the strong and repetitive chemical bonding among the molecules results in a large decrease in the internal energy term E, and therefore there is a net decrease in the free energy F, even though the magnitude of the entropy term has increased.

What is the connection between the symmetry of a crystal and its strong internal energy or binding energy? What happens is that, if molecules in some part of the crystal have found (by trial and error) some lowest-internal-energy (i.e., highest-binding-energy) configuration, other parts of the crystal will also hit upon the same configuration. Any other arrangement will result in dangling bonds etc., apart from being unlikely anyway. Thus the second law ensures order and symmetry, and is therefore the *primary* organizing principle for open systems.

The second law is all-pervasive, and everything that happens does so because the second law allows it. This is too sweeping a statement (although factually correct), and one would like to identify other, secondary, organizing principles. As argued in this book, symmetry is one such secondary organizing principle. And this is an important conclusion indeed. We have seen in the previous chapter that even the symmetry observed in a variety of complex networks can be rationalized in terms of a generalized version of the second law, with appropriate meaning attached to the term 'energy'. For example, in social networks, social cohesion is the equivalent of chemical cohesive energy in a crystal.

The similar-linkage-pattern (SLP) model of symmetry in complex networks is an example of how similarity of bonding, whether in crystals or in social, biological, or cultural networks, is responsible for the emergence of symmetry.

In keeping with the bottom-up approach to symmetry highlighted in this book, I have stated a *symmetry composition principle*. Symmetry often implies the existence of two or more equal or equivalent parts, namely the building blocks. The principle determines how any such observed symmetry can be understood as having arisen from the superposition of three contributing symmetries: (i) *intrinsic symmetry* (if any) of the building block; (ii) *placement symmetry* (if any), which is a description of how the building blocks have been placed with respect to one another; and (iii) *latent symmetry* (if any) manifesting itself under certain just-right conditions.

Latent symmetry can possibly become manifest when there are two or more equal or equivalent subunits involved. Under the right conditions (i.e., in the presence of certain specific placement symmetries), latent symmetry can become manifest as an *unexpected* property. The emergence of unexpected properties is also an important feature of complex systems, and many of them are indeed made up of equivalent subunits. Therefore, it is entirely possible that the symmetry-related emergences in complex systems may be *sometimes* because of the unexpected manifestations of latent symmetry. This needs to be investigated.

I came up with the idea of latent symmetry as far back as in 1987. But it has not found many applications yet. In fact, one reason why I have written this book is to highlight the possibility of latent symmetry manifesting itself in all sorts of natural phenomena under the right conditions. It is my hope that readers of this book will become alert to the possible existence of latent symmetry. The progress has been slow so far. Litvin's partition theorem is all we have at present for detecting the presence of latent symmetry.

Here is what Litvin's theorem achieves: Consider an object A having a symmetry H, from which a composite has been constructed by a group G of isometries. The theorem provides a sufficient condition that an isometry is a symmetry of this composite.

Vlachavas's work is an example of the kind of mistakes that can creep in if one is not aware of the possibility of latent symmetry making its presence felt. There is a clear need for taking latent symmetry to people who deal with symmetry. At present it is more a case of a *solution looking for a problem*.

Symmetry and broken symmetry are crucial to the way research in fundamental physics is carried out. Looking for manifestations of the spontaneous breaking of yet-unknown types of symmetry holds the key to answering some of the very basic questions about the origin of our universe, and about the quantization and unification of all the fundamental interactions of Nature, including the gravitational interaction. Gauge symmetry is the stuff natural laws are made of, and it is a *local* symmetry. A global symmetry is just a special case of the underlying gauge symmetry. Gauge symmetry is an intense field of study at present, particularly because it may hold the key for a unification of the gravitational interaction with the other three fundamental interactions, and for making breakthroughs in understanding the nature of quantum gravity.

Students of science should be introduced to symmetry at an early stage. At the very fundamental level, science is all about symmetry.

Bibliography

Achlioptas, D., R. M. D'Souza and J. Spencer (13 March 2009). 'Explosive percolation in random networks.' *Science*, 323: 1453.

Albert, R., H. Jeong and A.-L. Barabási (2000). 'Error and attack tolerance of complex networks.' *Nature*, 406: 378.

Alon, U. (26 September 2003). 'Biological networks: The tinkerer as an engineer'. *Science*, 301: 1866.

Alon, U. (2007). *An Introduction to Systems Biology: Design Principles of Biological Circuits*. London: Chapman and Hall / CRC.

Alspach, B., E. Dobson and J. Morris (2008). 'Symmetries of graphs and networks.'
http://www.birs.ca/workshops//2008/08w5047/report08w5047.pdf

Alvarez, S. and J. Echeverria (2010). 'New perspectives on polyhedral molecules and their crystal structures.' *Journal of Physical Organic Chemistry*, 23: 1080.

Ananthaswamy, A. (7 August 2010). 'The end of space-time.' *New Scientist*, 207: 28.

Anderson, P. W. (4 August 1972). 'More is different.' *Science*, 177: 393.

Anderson, P. W. (1981). 'Some general thoughts about broken symmetry.' In N. Boccara (Ed.) (1981). *Symmetries and Broken Symmetries in Condensed-Matter Physics*. Paris: IDSET.

Anderson, P. W. (1984). *Basic Notions of Condensed Matter Physics*. California: Addison-Wesley.

Andre, I., C. E. M. Strauss, D. B. Kaplan, P. Bradley and D. Baker (2008). 'Emergence of symmetry in homooligomeric biological assemblies.' *PNAS*, 105: 16148.

Bak, P. (1996). *How Nature Works: The Science of Self-Organized Criticality*. New York: Springer.

Ball, P. (1999a). *The Self-Made Tapestry: Pattern Formation in Nature*. Oxford: Oxford University Press.

Ball. P. (2 December 1999b). 'Transitions still to be made.' *Nature*, 402 (Supplement): C73.

Banados, M. and Reyes, I. (31 August 2017). 'A short review on Noether's theorems, gauge symmetries and boundary terms'. https://arxiv.org/pdf/1601.03616.pdf

Barabási, A,-L. (2003). *Linked: How Everything is Connected to Everything Else and What It Means for Business, Science, and Everyday Life*. Cambridge, MA: Perseus.

Barabási, A.-L. (24 July 2009). 'Scale-free networks: A decade and beyond.' *Science*, 325: 412.

Barabási, A.-L. and R. Albert (15 October 1999). 'Emergence of scaling in random networks.' *Science*, 286: 509.

Bianconi, G., P. Pin and M. Marsili (2009). 'Assessing the relevance of node features for network structure.' *PNAS*, 106: 11433.

Blinc, R. and B. Zeks (1974). *Soft Modes in Ferroelectrics and Antiferroelectrics*. Amsterdam: North-Holland.

Boccara, N. (Ed.) (1981). *Symmetries and Broken Symmetries in Condensed-Matter Physics*. Paris: IDSET.

Borges, R. M. (2000). 'How asymmetrical before it's asymmetrical?' *Journal of Biosciences*, 25(2): 121.

Brading, K. and E. Castellani (Eds.) (4 December 2003). *Symmetries in Physics: Philosophical Reflections*. Cambridge University Press.

Brandmuller, J. (1986). 'An extension of the Neumann-Minnigerode-Curie principle.' *Computation and Mathematics with Applications*, 12B: 97.

Browne, R. (15 December 2011). 'Noether's theorem: Uses and abuses'. https://inside.mines.edu/~tohno/teaching/PH505_2011/Ryan_FinalPaperNoetherThm.pdf

Burns, G. and A. M. Glazer (1990). *Space Groups for Solid State Scientists*. Boston: Academic Press.

Caldarelli, G. (2007). *Scale-Free Networks: Complex Webs in Nature and Technology*. Oxford: Oxford University Press.

Caldarelli, G., A. Capocci, P. de Los Rios and M. A. Munoz (2002). 'Scale-free networks from varying vertex intrinsic fitness.' *Physical Review Letters*, 89: 258702.

Carroll, Sean (2005). 'Hidden symmetries.' http://blogs.discovermagazine.com/cosmicvariance/2005/10/24/hidden-symmetries/

Castellano, C., S. Fortunato and V. Loreto (2007). 'Statistical physics of social dynamics.' http://arxiv.org/abs/0710.3256v1

Chaikin, P. M. and T. C. Lubensky (1995). *Principles of Condensed Matter Physics*. Cambridge: Cambridge University Press.

Chiba, T. and H. Nagahama (2001). 'Curie symmetry principle in nonlinear functional systems.' *Forma*: 16: 225.

Chua, L. O. (2005). 'Local activity is the origin of complexity'. *Inter. J. Bifurcat. Chaos*, 15(11), 3435–3456.

Countryman, D. R., J. M. Carney and J. L. Welsh, Jr. (1969). In A. G. H. Dietz (Ed.), *Composite Engineering Materials*. Cambridge Massachusetts: MIT Press.

Curie, P. (1884a). 'Sur la symetrie.' *Soc. Mineralog. France Bull. Paris*, 7: 418.

Curie, P. (1894b). 'Sur la symetrie dans les phenomenes physiques, symetrie d'un champ electrique et d'un champ magnetique.' *Journal de Physique*, 3: 393.

Dawkins, R. (1996). *Climbing Mount Improbable*. London: Viking.

Endress, P. K. (November 1999). 'Symmetry in flowers: Diversity and evolution.' *International Journal of Plant Science*, 160(S6): S3.

http://www.ncbi.nlm.nih.gov/pubmed/10572019

Enquist, M. and R. A. Johnstone (1997). 'Generalization and the evolution of symmetry preferences.' *Proceedings of the Royal Society of London B*, 264: 1345.

Feynman, R. P. (1965). *The Character of Physical Law*. Penguin Books.

Feynman, R. P. (1985). *QED*. Penguin Books.

Fichtner, K. (1986). 'Non-space-group symmetry in crystallography.' *Computation and Mathematics with Applications*, 12B: 751.

Flack, H. D. (2003). 'Chiral and achiral crystal structures.' *Helvetica Chimica Acta*, 86: 905.

Fousek, J. and L. E. Cross (2001). *Ferroelectrics*, 252: 171.

Fousek, J., D. B. Litvin and L. E. Cross (2001). 'Domain geometry engineering and domain average engineering of ferroics.' *Journal of Physics: Condensed Matter*, 13: L33.

Garcia-Bellido, A. (1996). 'Symmetries throughout organic evolution.' *PNAS*, 93: 14229.

Garlaschelli, D. and M. I. Loffredo (2008). 'Maximum likelihood: Extracting unbiased information from complex networks.' *Physical Review E*, 78: 015101.

Garlaschelli, D., F. Ruzzenenti and R. Basosi (2010). 'Complex networks and symmetry I: A review.' *Symmetry*, 2: xxx. arXiv:10009.4489v1 [q-fin.GN]

Garrido, A. (2010). 'A survey on complex networks.' *BRAIN. Broad Research in Artificial Intelligence and Neuroscience*, 2(1): 63.

Garrido, A. (2011). 'Symmetry in complex networks.' *Symmetry*, 3: 1.

Goldberg, D. (2014). *Universe in the Rearview Mirror: How Hidden Symmetries Shape Reality*. Plume.

Goldstone, J. (1961). *Nuovo Cimento*, 19: 154.

Godsil, C. and G. Royle (2001). *Algebraic Graph Theory*. Berlin: Springer.

Goodsell, D. S. and A. J. Olson (2000). 'Structural symmetry and protein function.' *Annual Reviews of Biophysics and Biomolecular Structure*, 29: 105.

Gross, D. J. (1996). 'The role of symmetry in fundamental physics.' *PNAS*, 93: 14256.

Hawking, S. W. and T. Hertog (2002). 'Why does inflation start at the top of the hill?' *Phys. Rev.*, D66: 123509. www.arXiv:hep-th/0204212.

Hawking, S. W. and T. Hertog (10 February 2006). 'Populating the landscape: A top down approach'. www.arXiv:hep-th/0602091

Hawking, S. and L. Mlodinow (2010). *The Grand Design*. New York: Bantam Books.

Hermann, C. (1934). *Z. Kristallogr.*, 89: 32.

Holme, P. (3 May 2006a). 'Detecting degree symmetries in networks.' arXiv/physics/0605029v1.

Holme, P. (2006b). 'Local symmetries in complex networks.' *Physical Review E*, 74: 036107.

Hua, D., X. Yanghua, W. Wei, J. Li and X. Momiao (2008). 'Symmetry of metabolic network.' *Journal of Computer Science and Systems Biology*, 1: 1.

Ismael, J. (1997). 'Curie's principle.' *Synthese*, 110: 167.

Kisak, P. F. (Ed.) (15 May 2016). *Symmetry Breaking & Symmetry in Cosmology: The Fundamentals of the Cosmological Timeline*. CreateSpace Independent Publishing Platform, SC, USA.

Koptsik, V. A. (1983). 'Symmetry principle in physics.' *Journal of Physics C: Solid State Physics*, 16: 23.

Krauss, L. (2017). *The Greatest Story Ever Told . . . So Far*. Simon & Schuster, UK.

Landau, L. D. (1937). *Phys. Z. Sowjet.* 11: 26 (in Russian). For an English translation, see D. ter Haar (ed.) (1965), *Collected Papers of L. D. Landau.* New York: Gordon and Breach.

Lederman, L. M. and C. T. Hill (2004/2008). *Symmetry and the Beautiful Universe.* New York: Prometheus Books.

Levy, Y., S. S. Cho, T. Shen, J. N. Onuchic and P. G. Wolynes (2005). 'Symmetry and frustration in protein energy landscapes: a near degeneracy resolves the Rop dimer mystery.' *PNAS*, 102: 2373.

Li, H., R. Helling, C. Tang and N. Wingreen (1996). 'Emergence of preferred structures in a simple model of protein folding.' *Science*, 273: 666.

Lin, S. K. (1996a). 'Correlation of entropy with similarity and symmetry.' *Journal of Chemical Information and Computer Sciences*, 36: 367. http://www.mdpi.org/lin/similarity.pdf

Lin, S. K. (1996b). 'Molecular diversity assessment: Logarithmic relations of information and species diversity and logarithmic relations of entropy and indistinguishability after rejection of Gibbs paradox of entropy of mixing.' *Molecules*, 1: 57.

Lin, S. K. (1996c). 'Gibbs paradox of entropy mixing: Experimental facts, its rejection, and the theoretical consequences.' *Journal of Theoretical Chemistry*, 1: 135.

Lin, S. K. (1999a). 'Ugly symmetry, beautiful diversity.' http://www.mdpi.org/lin/uglysymnews.htm

Lin, S. K. (1999b). 'Diversity and entropy.' *Entropy*, 1:1. http://www.mdpi.org/entropy/papers/e1010001.pdf

Lin, S. K. (2001). 'The nature of the chemical process: 1. Symmetry evolution - Revised information theory, similarity principle and ugly symmetry.' *International Journal of Molecular Sciences*, 2: 10. http://arxiv.org/ftp/physics/papers/0105/0105024.pdf

Lin, S. K. (2008). 'Gibbs paradox and the concepts of information, symmetry, similarity and their relationship.' *Entropy*, 10: 1.

Litvin, D. B. and V. K. Wadhawan (2001). 'Latent symmetry and its group theoretical determination.' *Acta Crystallographica*, A57: 435.

Litvin, D. B. and V. K. Wadhawan (2002). 'Latent symmetry.' *Acta Crystallographica*, A58: 75.

Litvin, D. B., V. K. Wadhawan and D. M. Hatch (2003). 'Latent symmetry and domain average engineered ferroics' *Ferroelectrics*, 292: 65.

MacArthur, B. D. and J. W. Anderson (2006). 'Symmetry and self-organization in complex systems.' e-print arXiv:cond-mat/0609274.

MacArthur, B. D., R. J. Sánchez-García and J. W. Anderson (2007). 'On automorphism groups of networks.' e-print arXiv:0705.3215v2 [physics.soc-ph].

MacArthur, B. D., R. J. Sánchez-García and J. W. Anderson (2008). 'Symmetry in complex networks.' *Discrete Applied Mathematics*, 156: 3525.

MacArthur, B. D. and R. J. Sánchez-García (2009). 'Spectral characteristics of network redundancy.' *Physical Review E*, 80: 026117.

Mainzer, K. (2005). *Symmetry and Complexity: The Spirit and Beauty of Nonlinear Science*. Singapore: World Scientific.

Mainzer, K. (2014). 'The cause of complexity in Nature: An analytical and computational approach'. In I. Zelinka *et al.* (eds.), *How Nature Works*, Emergence, Complexity and Computation 5, DOI: 10.1007/978-3-319-00254-5_2. Springer International Publishing, Switzerland (2014).

Mainzer, K. and L. Chua (20 March 2013). *Local Activity Principle: The Cause of Complexity and Symmetry Breaking*. Imperial College Press.

Milo, R., S. Shen-Orr, S. Itzkovitz, N. Kashtan, D. Chklovskii and U. Alon (2002). 'Network motifs: Simple building blocks of complex networks'. *Science*, 298: 824.

Minnigerode, B. (1884). 'Untersuchungen uber die Symmetrieverhaltnisse und die Elasticitat der Krystalle.' *Nachr. Akad. Wiss. Gottingen, Math-phys. / Klasse II a* 184: 195.

Monod, J., J. Wyman and J. P. Changeux (1965). 'On the nature of allosteric transitions: A plausible model.' *Journal of Molecular Biology*, 12: 88.

Nakamura, N. and H. Nagahama (2000). 'Curie symmetry principle: Does it constrain the analysis of structural geology?' *Forma*, 15: 87.

Newman, M. E. J. (2003). 'The structure and function of complex networks.' http://arxiv.org/abs/cond-mat/0303516v1

Noether, E. (1918). 'Invariante Variationsprobleme'. *Nachr. d. Konig. Gesellsch. d. Wiss. zu Gottingen, Math-phys. Klasse*, 1918: 235-257. [*Transp. Theory Statist. Phys.*, 1,186 (1971)]. Also see Tavel, M. A. (1971), 'Milestones in mathematical physics: Noether's theorem.' *Transp. Theory Statist. Phys.*: 1 183–185.

Nye, J. F. (1976). *Physical Properties of Crystals*. Oxford: The Clarendon Press.

Palmer, A. R. (2003). 'Eclectic reflections on biological asymmetry.' http://www.biology.ualberta.ca/palmer/asym/asymmetry.htm

Palmer, A. R. (2004). 'Symmetry breaking and the evolution of development.' *Science*, 306: 828.

Park, J. and M. E. J. Newman (2003). 'Origin of degree correlations in the Internet and other networks.' *Physical Review E*, 68: 026112.

Park, J. and M. E. J. Newman (2004). 'Statistical mechanics of networks.' *Physical Review E*, 70: 066117.

Petitjean, M. (2003). 'Chirality and symmetry measures: A transdisciplinary review.' *Entropy*, 5: 271.

Petitjean, M. (2007). 'A definition of symmetry.' *Symmetry: Culture and Science*, 18: 99.

Pond, R. C. and D. S. Vlachavas (1983). *Proceedings of the Royal Society of London*, A386: 95.

Prigogine, I. (1977). 'Time, structure and fluctuations.' Nobel lecture.

http://nobelprize.org/nobel_prizes/chemistry/laureates/1977/prigogine-lecture.pdf

Ratsaby, J. (2008). 'An algorithmic complexity interpretation of Lin's third law of information theory.' *Entropy*, 10:6.

Rosen, J. (1995). *Symmetry in Science. An Introduction to the General Theory*. Berlin: Springer-Verlag.

Rosen, J. (2008). *Symmetry Rules: How Science and Nature are Founded on Symmetry*. Berlin: Springer-Verlag.

Ruzzenenti, F., D. Garlaschelli and R. Basosi (2010). 'Complex networks and symmetry II: Reciprocity and evolution of world trade.' *Symmetry*, 2: 1710.

Saltus, R. (August 2007). 'Broken symmetry.' HHMI Bulletin, 20(3): 1. http://www.hhmi.org/bulletin/aug2007/pdf/Symmetry.pdf

Schwichtenberg, J. (25 June 2015). *Physics from Symmetry* (Undergraduate Lecture Notes in Physics). Springer Nature.

Sethna, J. P. (1992). 'Order parameters, broken symmetries, and topology.' In L. Nagel and D. Stein (Eds.), *1991 Lectures in Complex Systems*. Santa Fe Institute Studies in the Sciences of Complexity. Proc. Vol. XV. Addison-Wesley.

Sethna, J. P. And M. Huang (1992). 'Meissner effects and constraints.' In L. Nagel and D. Stein (Eds.), *1991 Lectures in Complex Systems*. Santa Fe Institute Studies in the Sciences of Complexity. Proc. Vol. XV. Addison-Wesley.

Sheftal, N. N. (1966a). 'The definition of symmetry.' In A. V. Shubnikov and N. N. Sheftal (Eds.), *Growth of Crystals*, Vol. 4. New York: Consultants Bureau.

Sheftal, N. N. (1966b). 'The physical meaning of symmetry.' In A. V. Shubnikov and N. N. Sheftal (Eds.), *Growth of Crystals*, Vol. 4. New York: Consultants Bureau.

Sheftal, N. N. (1976). 'A crystal as a medium that order phenomena.' In N. N. Sheftal (Ed.), *Growth of Crystals*, Vol. 10. New York: Consultants

Bureau.

Shen-Orr, S., R. Milo, S. Mangan and U. Alon (2002). 'Network motifs in the transcriptional regulation network of Escherichia coli.' *Nature Genetics*, 31:64.

Shubnikov, A. V. (1944). 'Advances in the study and application of symmetry.' General Meeting of the Academy of Sciences of the USSR, October 14-17, 1944. *Izd. Akad. Nauk SSSR* (1945), p. 24.

Shubnikov, A. V. and V. A. Koptsik (1974). *Symmetry in Science and Art*. New York: Plenum Press.

Sirotin, Yu. I. and M. P. Shaskolskaya (1982). *Fundamentals of Crystal Physics*. Moscow: Mir Publishers.

Soderberg, B. (2002). *Physical Review E*, 66: 066121

Stanley, H. E. (1999). 'Scaling, universality, and renormalization: Three pillars of modern critical phenomena.' *Reviews of Modern Physics*, 71: S358.

Stavsky, Y. and N. J. Hoff (1969). In A. G. H. Dietz (Ed.), *Composite Engineering Materials*. Cambridge Massachusetts: MIT Press.

Sundermeyer, K. (2014). *Symmetries in Fundamental Physics (Fundamental Theories of Physics)*. Springer; softcover reprint of the original 2014 edition (22 September 2016).

Sweatman, W. (14 June 2016). 'Symmetry for dummies: Noether's theorem'. https://hackaday.com/2016/06/14/symmetry-for-dummies-noethers-theorem/

Trodden, M. (2005). 'The search for new particles and symmetries.' http://inpa.lbl.gov/pbar/talks/S9_Trodden.pdf

Vainshtein, B. K. and A. A. Chernov (Eds.) (1988). *Modern Crystallography*. New York: Nova Science Publishers.

Vlachavas, D. S. (1984). *Acta Crystallographica A*, 40: 213.

Wadhawan, V. K. (1987). 'The generalized Curie principle, the Hermann theorem, and the symmetry of macroscopic tensor properties of composites.' *Materials Research Bulletin*, 22: 651.

Wadhawan, V. K. (1998). 'Towards a rigorous definition of ferroic phase transitions.' *Phase Transitions,* 64: 165.

Wadhawan, V. K. (2000). *Introduction to Ferroic Materials.* Amsterdam: Gordon and Breach.

Wadhawan, V. K. (July 2002). 'Ferroic materials: A primer.' *Resonance*, p. 15.

Wadhawan, V. K. (2007). *Smart Structures: Blurring the Distinction Between the Living and the Nonliving.* Oxford: Oxford University Press.

Wadhawan, V. K. (2010). *Complexity Science: Tackling the Difficult Questions We ask about Ourselves and about Our Universe.* Saarbrücken: LAP Lambert Academic Publishing.

Wadhawan, V. K. (2017/2018). *Understanding Natural Phenomena: Self-Organization and Emergence in Complex Systems*. CreateSpace Independent Publishing Platform, USA.

Wang, T., J. Miller, N. S. Wingreen, C. Tang and K. A. Dill (2000). 'Symmetry and designability of lattice protein models.' *Journal of Chemical Physics*, 113: 8329.

Wang, Yan and Xiao (2009). 'Symmetry in world trade network.' *J. Syst. Sci. Complex*, 22: 280.

Wasserman, S. and K. Faust (1994). *Social Network Analysis: Methods and Applications*. New York: Cambridge University Press.

Watts, D. J. and S. H. Strogatz (1998). 'Collective dynamics of "small world" networks.' *Nature*, 393: 440.

Wolynes, P. G. (1996). 'Symmetry and the energy landscapes of biomolecules.' *PNAS*, 93: 14249.

Xiao, Y., W-T. Wu, H. Wang, M. Xiong and W. Wang (2008a). 'Symmetry-based structure entropy of complex networks.' *Physica A*, 387: 2611.

Xiao, Y., M. Xiong, W. Wang and H. Wang (2008b). 'Emergence of symmetry in complex networks.' *Physical Review E*, 77: 066108.

Xiao, Y., B. D. MacArthur, H. Wang and M. Xiong (2008c). 'Network quotients: structural skeletons of complex systems.' *Physical Review E*, 78: 046102.

Zee, A. and R. Penrose (2016). *Fearful Symmetry – The Search for Beauty in Modern Physics* (Princeton Science Library). Princeton University Press.

Index

2-dimensional higher group, 34
2-group, 34
Abelian group, 22
acoustic phonons, 104
action function, 89
action symmetry, 88
analogy and classification are
    symmetry, 11
aperiodic crystal, 158
apparent laws of physics, 124
approximate symmetry, 61
approximate symmetry of graphs,
    61
asymmetric unit, 59, 131
asymptotic freedom, 118
automorphism, 30, 43, 61, 62, 65
automorphism group, 43, 44, 64
automorphism partition, 45, 154

BA model of networks, 63, 157,
    158
Barabási-Albert model, 64
baryons, 118
base point, 33
basis, 31
basis vectors, 26
Big Bang, 119, 124, 125
bijection, 30
bijective function, 30
bijective morphism, 30
bilateral symmetry, 4
bipartite graph, 40, 157

bosons, 113
bottom-up approach to
    symmetry, 135
broken (or 'hidden') new
    symmetry, 111
broken symmetry, xiii, 1, 6, 20,
    93, 94, 100, 101, 102, 103,
    106, 111, 122

categorical groups, 34
category, 29, 30, 31
category of groups, 34
Cayley graph, 44
charge-reversal symmetry, 83
class, 30
cluster, 41
clustering coefficient, 42
clustering or transitivity, 42
common ancestor, 70
compact Lie group, 32
complete bipartite graph, 40
complex systems, 1, 93
conjugate transpose, 28
connected graph, 41
connected subgraph, 41
connected triple, 42
continuous broken symmetries,
    99
continuous groups, 27
coset decomposition of a group,
    23
coset representative, 23

cosmic inflation, 119
CPT symmetry, 106
crossed module, 35
cross-ply plywood, 78
Curie principle, 71, 96
Curie-Shubnikov principle, 73, 75

degree coefficient, 156
degree of complexity, 137
degree symmetry, 155, 156, 160
degree theorem, 40
degree distribution, 156
degree symmetry, 155
dextral asymmetry, 107
diameter of a graph, 41
differentiable or differential symmetry, 88
directed graph, 39
disconnected graph, 41
discrete broken symmetries, 105
discrete groups, 21, 34
dissymmetrization, 77
dissymmetry, 72
distinction between potential symmetry and latent symmetry, 132
domain states, 144
domain-average engineered (DAE) ferroic, 145
dynamic symmetry, 20

electromagnetic interaction, 112, 114
electroweak interaction, 116, 117
embracing group, 71

emergence of symmetry in real-life networks, 158
emergent properties, 152
emergent symmetry, 7, 135, 152
epimorphism, 29
equal or identical placement of equal parts, 49, 59, 76
evolution of improved evolvability, 56
evolutionary underpinning of symmetry in biological forms, 55

factor group, 24, 25
fermions, 113
ferroic phase transitions (FPTs), 97
field, 28, 84
finite Lie groups, 88
force field, 114
formal definition of latent symmetry, 139
four types of forces or interactions, 111
function, 84
fundamental group, 33, 35, 71
fundamental theorem of symmetry, xiv, 133

gauge bosons, 114, 119, 120
gauge group, 85
gauge invariance, 84, 85, 115, 116, 119
gauge invariance theory, 82
gauge symmetry, 5, 82, 84, 86, 104, 114, 116, 119, 121, 165
gauge symmetry of QED, 117

gauge transformation, 82, 115

gauge-symmetry groups, 84

geodesic path, 41

geometrical symmetry, 10

geometrization of symmetry, 47

global symmetry, 81, 83, 86, 88, 149

gluons, 86

Goldstone's theorem, 100

good get richer mechanism, 66

grand unification theories (GUTs), 118

graph, 39

gravitational interaction, 102, 111-114, 117-119, 123, 125, 165

graviton, 120

group objects, 34

groupal categories, 34

groupoid, 30, 31, 149

growth of a crystal as an ordering process, 47

growth with preferential attachment, 2, 49

GUT scalar field, 119

hadrons, 120

handshaking lemma, 40

Hermann theorem of crystal physics, 77

Hermitian adjoint, 28

hexply, 79

Higgs boson, 102, 120

Higgs field, 102, 119, 120

homomorphism, 29

hubs, 63

improper rotations, 28

infinite-dimensional Lie groups, 88

initial degree, 66, 157, 158

invariance group, 140, 149

invariance under permutation of 'external' properties, 66

isometry, 139

isomorphic, 30

isomorphism, 30

isotopic spin, 116

kaleidoscopic embryologies, 55

labelled graph, 64

Lagrange theorem for subgroups, 25

Landau theory of phase transitions, 95

large category, 30

latent symmetry, xiii, 53, 55, 129, 130, 133, 134, 138-140, 145, 149

latent symmetry and algorithmic information, 137

latent symmetry and domain-average engineered ferroic materials, 144

latent symmetry and potential symmetry, 129

latent symmetry in complex networks, 151

left-right symmetry, 54

leptons, 120

Lie groups, 32

Litvin's partition theorem for latent symmetry, 140

local-activity principle, 108
local symmetry, 47
local transformations, 81
Lorentz group, 85

manifold, 29
mapping, 84
mathematical networks, 39
maximalistic use of the
    symmetry principle, 15
measures of symmetry of
    networks, 154
mesons, 118
Mexican-hat potential, 101
minimalistic use of the symmetry
    principle, 15
module, 34
monoids, 30
monomorphism, 29
morphisms, 29
mother of all organizing
    principles, 163
M-theory, 124, 126, 127

Nambu-Goldstone modes, 100
Nambu-Goldstone theorem, 102
neighbourhood of a point, 29
network, 39
network motifs (NMs), 65
Neumann principle, 73
Neumann-Minnigerode-Curie
    (NMC) principle, 76
Neumann-Minnigerode-Curie-
    Shubnikov principle, 76
n-groups, 34
no-boundary condition, 126
Noether's first theorem, 88

Noether's second theorem, 91
Noether's theorems, 86
non-space-group symmetry, 57
normal subgroup, 35
null model, 156

object, 29
observables, 83, 113
OD crystal structures, 149
open sets, 28, 29
optical indicatrix, 77
orbit, 45
order parameter, 93, 96
organizing principle, 49, 50
orientation states, 144
origins of symmetry in complex
    networks, 156
orthogonal group, 27
orthogonal group O(3), 27
orthogonal matrices, 27

partial function, 31
partial symmetry, 149
partition, 44
path, 33
p-branes, 124
permutation, 27, 43, 61
permutation group, 27, 150
permutation symmetry, 10, 43,
    69, 152
permutation symmetry in graphs
    and networks, 43
permutational and more general
    symmetries of graphs, 60
phase transition, 93
photoelastic tensor, 80

placement symmetry, 2, 7, 59, 69, 129, 132, 133, 136, 139, 149

Planck time, 119

point field, 27, 84

pointed space, 33

polytypes, 149

potential symmetry, 4, 131, 132, 153

predictability is symmetry, 14

preferential attachment, 52, 64, 157

primary organizing principle, 4, 160, 164

principle of least action, 89

principle of local activity, 109

proper class, 30

proper rotations, 28

prototype symmetry, 98

quantum chromodynamics (QCD), 86, 118

quantum electrodynamics (QED), 114

quantum gravity, 118

quarks, 118

quasi-isolated systems, 12

quotient group, 24, 25, 35

random asymmetry, 107

random graph or network, 42

real-life complex networks, 45, 46

reciprocity relations, 17

reduction is symmetry, 11

regular graph, 40, 62

renormalization group, 63

renormalization procedures, 115

reproducibility is symmetry, 13

rich getting richer, 63

robustness of real networks, 64

role of placement symmetry in revealing latent symmetry, 148

rotation-inversion group, 28

r-regular graph, 40

scale invariance, 61, 63

scale-free networks, 46, 63

scale-invariance symmetry, 63

secondary organizing principle, 4, 156, 164

segmentation, 54

semigroup, 30, 31, 149

semi-symmetric graph, 44

similar linkage pattern (SLP), 2, 49, 52, 64, 152, 153, 154, 156, 158

similar linkage patterns and symmetry, 49

similar-linkage-pattern (SLP) model of symmetry, 157, 164

simple graph, 39

simple group, 24, 35

simple Lie group, 32, 36, 37

simple, or simplifiable, systems, 12

simply connected group, 35, 36

simply connected space, 33

sinistral asymmetry, 107

site symmetry, 59

small category, 30

small class, 30

small-world effect, 62

social networks and the meaning of cohesive energy, 160

spanning subgraph, 41
special orthogonal group, SO(n), 28
special unitary group, SU(n), 27, 28
speciation, 70
spontaneous breaking of symmetry, 94
static symmetry, 20
statistical equivalence of vertices, 65
stochastic notion of symmetry, 61
stochastic symmetry, 69
stochastically symmetric graph ensemble, 62
strict 2-group, 34
string theory, 123
strong nuclear interaction, 112
structural equivalence in graph theory, 64
structural redundancy, 65
supergravity, 103, 122, 123
supersymmetry, 103, 122, 123
symmetric bicliques, 156-158
symmetric derivative, 88
symmetrization, 75, 76, 77
symmetry and the conservation laws of physics, 87
symmetry and the folding efficiency of proteins, 159
symmetry as a secondary organizing principle, 4, 50, 52
symmetry compensation law, 98
symmetry composition principle, xiii, 6, 7, 133, 134
symmetry element, 9

symmetry enhancement, 77
symmetry evolution principle, 15
symmetry evolved by stalked jellyfishes, 54
symmetry group of a crystal, 25
symmetry in homooligomeric biological assemblies, 56
symmetry in real-life networks, 62
symmetry operation, 9
symmetry principle, 15
symmetry transformations, 25, 59
symmetry-breaking transitions, 116
symmetry conservation principle, 5
system, 45

topological group, 34
topological space, 28
topological transformation, 61
translational symmetry, 25
transverse isotropy, 77, 79, 80

ugly symmetry, 17
uncertainty principle, 118
underlying group, 35
undirected graph, 39
unit cell, 26
unit interval, 33
unitary group U(n), 28
unitary matrix U, 28
universal covering of G, 35

virtual group, 31
Vlachavas's theorems, 147

weak nuclear interaction, 112

Yang-Mills field, 116
Yang-Mills gauge symmetry, 117
Yang-Mills theory, 117

Acknowledgements

The writing of this monograph was started during my tenure as a Raja Ramanna Fellow at the Bhabha Atomic Research Centre (BARC), Mumbai. I am grateful to Dr. Srikumar Banerjee, Chairman, Atomic Energy Commission, Government of India, for his consistent support during my career in the Department of Atomic Energy (DAE). I also wish to thank Dr. S. Kailas, Director, Physics Group, BARC, for all his help and support.

My student Dr. Indranil Bhaumik helped me substantially with the library work. Dr. A. K. Rajarajan helped with some of the drawings.

Prof. Daniel Litvin was my collaborator on latent-symmetry studies. We published three papers together. The group-theoretical treatment of latent symmetry was largely his contribution. He is a high-calibre mathematician, and our interaction often made me realize that I as a physicist do not always have the rigour of logic that is so characteristic of mathematics. On a lighter note, I may mention that the celebrated Gödel theorem has indeed made things tough for many mathematicians! As Gregory Chaitin has suggested, mathematics should now be done more and more like physics.

Prof. Dorian Hatch joined me and Prof. Litvin in some of the work on latent symmetry, and I am thankful to him also for several useful discussions.

My association with symmetry goes back to the very first day of my career in the DAE as an X-ray crystallographer. That was in 1968. Since then I have made forays into several other areas of physics: ferroelasticity; phase transitions; ferroic materials; crystal growth; ferroelectric ceramics; crystal physics; the utilitarian role of symmetry in materials science; smart structures; and complex systems. Symmetry considerations pervade all of science. I am currently trying to fathom the intricacies of complexity science. Even here the exciting and surprising thing is that complex networks can be richly symmetric, as discussed in this book.

Prof. A. M. Glazer ('Mike Glazer' to everybody), a crystallographer par excellence (and much else!), was an early influence on me, ever since I spent a year in his group at the Clarendon Laboratory, Oxford, as a Nuffield Foundation Fellow. That was in 1979-80. Mike has been a great source of strength to me as a scientist. Thank you, Mike.

ABOUT THE AUTHOR

Vinod Wadhawan earned his M. Sc. degree in solid-state physics from the University of Delhi, and Ph. D. degree in X-ray crystallography from the University of Bombay. His areas of specialization include crystal physics, phase transitions, the utilitarian role of symmetry in materials science, ferroic materials, crystal growth, smart structures, and complex systems. A crystallographer and teacher, he served for two terms as member of the International Union for Crystallography Commission on Teaching. He has been a recipient of the prestigious Raja Ramanna Fellowship of the Department of Atomic Energy, Government of India, for working at the Bhabha Atomic Research Centre, Mumbai. Before that he served as 'Outstanding Scientist' and Head, Laser Materials Division, at the Raja Ramanna Centre for Advanced Technology, Indore. He was till 2011 an Associate Editor of *Phase Transitions* (Taylor & Francis), having done editorial work for this journal for a quarter of a century.

He is an avid blogger. His blog is called THE VINOD WADHAWAN BLOG: Celebrating the Spirit of Science and the Scientific Method.
http://vinodwadhawan.blogspot.com/

www.ingramcontent.com/pod-product-compliance
Lightning Source LLC
Chambersburg PA
CBHW071714170526
45165CB00005B/2012